HANDBOOK OF LOCAL AREA NETWORK SOFTWARE

HANDBOOK OF LOCAL AREA NETWORK SOFTWARE

Concepts and Technology

PAUL J. FORTIER

CRC

CRC Press
Boca Raton Ann Arbor Boston

Library of Congress Catalog Card Number 89-83758

Direct all inquiries to CRC Press, Inc. 2000 Corporate Blvd., N.W., Boca Raton, Florida 33431.

International Standard Book Number 0-8493-7404-9
Printed in the United States of America

90 91 92 93 10 9 8 7 6 5 4 3 2

PREFACE

Ever since the successful introduction of the first computer network, Arpanet, in the late 1960s, hundreds of other computer networks have come into being. These networks run the gamut from global world networks, national networks, metropolitan networks, down to local area networks of various sizes and complexity.

With the continuing expansion of the number of host computers (Dec announced the shipping of the 100,000 th VAX), workstations (Sun, Apollo), personal computers and terminals, and the attendant increase in processing and information demands, it can be seen how the role of the network (wide area or otherwise) has expanded and been solidified. As such we will see more and more of these computing devices being linked together into networks for the purpose of resource sharing, expanding availability of resources, information access, and better service to the expanding user base.

To meet the needs of this growing community Local Area Network vendors have developed new systems and software products providing a myriad of services before unimaginable.

Local area networks have provided a means for enterprises to expand their computing resources in a logical controlled fashion. These growing networks have given these institutions a means to more readily and effectively utilize their most important corporate resource; INFORMATION. This one element of networking has led to the current information explosion.

Because of the vastly expanding requirements for computing and information exchange and sharing, the demands for extended services from the networks has also skyrocketed. We are seeing another revolution in the computing and technology arena as that which occurred when operating systems first arrived. That is, ways to simplify and control the general forms of services required by all are migrating from user problems to systems developers (vendors) problems. The solutions need to be general and performed at the system services level.

This systems services software will take on many forms and address diverse computing problems. For example; information management, electronic mail, built-in boards, resource sharing, teleconferencing, design management, fault tolerance and configuration management to name a few. All of these software elements will add new dimensions to the LANs utility to users.

This book will be divided into two parts. The first will be an intensive section examining the various classes of software utilized in LANs in regards to their structure and operation. The second section will survey the available LAN software, and discuss concepts and applications thereof.

The emphasis of this book is to provide to computer scientists, engineers, network designers, network users, applications programmers and students, the underlying concepts used in the design, operation and use of viable computer communications networks. It provides these individuals with an understanding at all levels of a LANs architecture and uses in todays computing market.

CONTENTS

1

INTRODUCTION TO LOCAL AREA NETWORKS

Local area networks (LANs) continue to grow in popularity, as evidenced by the many new LAN products and the publicity related to them. Numerous classroom courses on the subject are offered as well as conferences and sessions dedicated to them. The draw is the promise of using LANs to interconnect various computers and resources into a unified system with more power and performance than is available from conventional nondistributed approaches.

Potential users of local area networks want to share information and programs, have increased computing power, or get at specified equipment. A LAN must provide services and interfaces compatible with a wide array of user requirements based on intended use. A LAN needs to be more than just a wire and connection mechanism. It must provide upper-level services to users that aid in their overall applications, design, development, and use—as do today's operating systems. This class of service is the next great development opportunity. The lowest-level communications equipment and protocols are standardized and readily available off the shelf, but the upper-layer software is not so readily available. There is great opportunity for the company that can provide products for the upper-layer needs.

How did we get where we are today? How did networks evolve? Basically, system designers have used four techniques to interconnect computer equipment. They are:

1. Centralized
2. Decentralized
3. Distributed
4. Networked

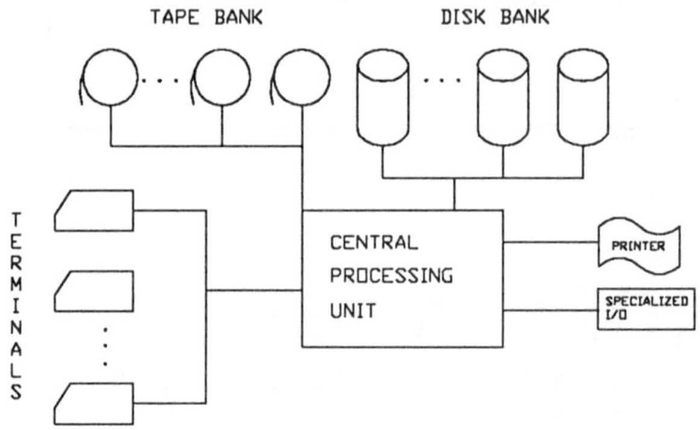

Figure 1-1 Centralized Computer Interconnect.

A centralized interconnect environment is a self-contained system capable of autonomous operation (Figure 1-1).

A centralized interconnect exhibits a master/slave relationship between the CPU and the attached devices. The attached devices are typically strung off the I/O bus of the computer. Data is transferred in data blocks via direct address, using the centralized computer system's operating system synchronization and timing mechanisms. It is engineered as a single, stand-alone entity where all devices are linked and controlled via the central processing subsystem.

A decentralized system is a computing environment where not all the devices (CPUs, I/O, disks, etc.) are at a single site (Figure 1-2). This class of computer interconnection typically uses serial communications media, connecting remote sites as if they were terminals. It uses a master/slave control relationship and ships data in blocks, as communications in peripheral devices are done in centralized computing systems.

Distributed interconnection is described as a collection of computers connected via communications links and unified by a systemwide operating system (Figure 1-3). This class of computer communications system is typified by a multiplicity of resources (physical and logical), a systemwide operating system, services requested by name not by location, and computing functions that are dispersed among several physical computing elements. The system is viewed and acts as a virtual uniprocessor.

A computer network interconnection system is defined as an interconnected collection of autonomous computers (Figure 1-4) that are interconnected for the purpose of exchanging information and services. Each computer has its own autonomous operating system, and there is no master/slave relationship. The components of such a system are cooperatively autonomous but mutually suspicious of each other. They use dedicated front-end processors, or special purpose input/output software and hardware for data communications, with the unit of transfer typically being the packet. The

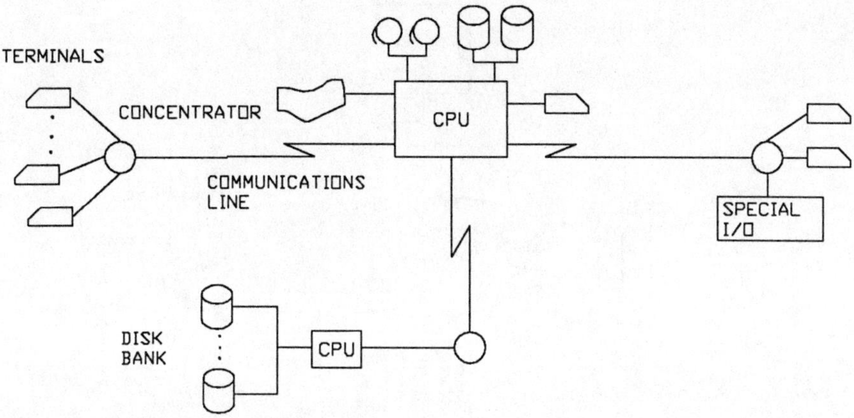

Figure 1-2. Decentralized Computer System.

front-end processors or interconnection hardware and software require handshaking-type protocols to effect communications.

This classification of interconnection schemes does not appear to leave a clean place to insert a local area network, and in the past this was the case. Local area networks fall in between the network classification and the distributed interconnect classification, due mainly to the level of integration of the LAN components into either loose couplings (network) or tight couplings (distributed). Computer networks and distributed systems have their roots in the early to late 1960s. Event A, which occurred in

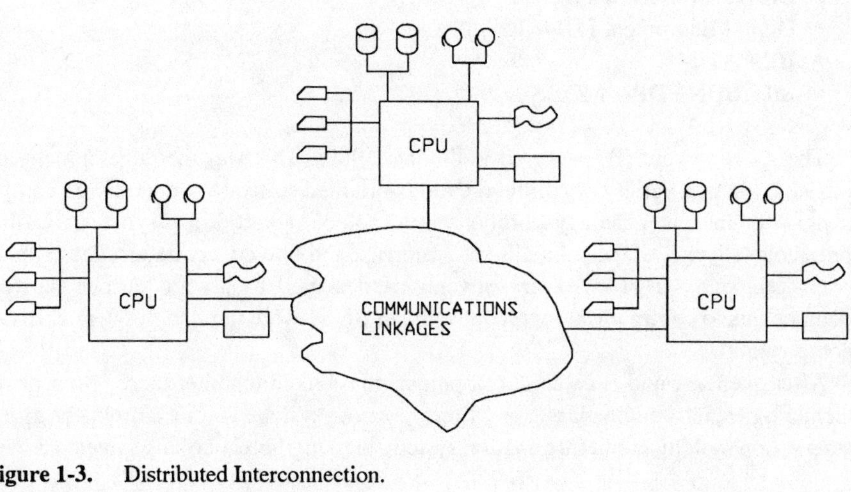

Figure 1-3. Distributed Interconnection.

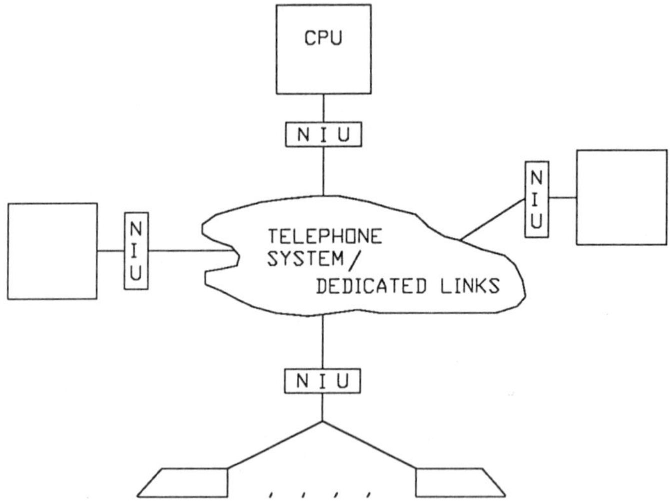

Figure 1-4. Network Interconnection System.

the early 1960s, was the introduction of time-sharing systems. These types of systems allowed multiple-terminal users to seemingly use the system's resources simultaneously. This feat was accomplished using quantum time management techniques. A few such early operating systems were:

- The IBM 360 RTOS Operating System
- Honeywell 600 Gecos III
- Univac 1108 Demand
- Digital Equipment PDP8 TSS/8
- IBM ATLAS
- MIT/BBN PDP-1 MODS

The second event (B) occurred in the late 1960s. This was the development and introduction of a variety of communications-oriented software and hardware components—for instance, the asynchronous line protocol (Start/Stop), synchronous line protocols (Bisync, SDLC), intelligent terminals, and line concentrators.These protocols defined a set of rules or conventions designed to permit various network components to communicate with one another (these were the first network software components).

After events A and B came developments that linked together these two separate technologies, introducing large, real-time operating systems—for example, the airline reservation system, credit information system, banking/hotel booking systems, inventory control, and point of sale systems. These represented the beginning of the evolution to information exchange networks.

The third major event (C) occurred in the early 1970s. This event was the dramatic drop in the price of computing assets such as minicomputers, microprocessors, peripheral devices, and communications costs. This translated directly into cheaper networking components.

The combination of events A, B, and C led to computer networking. Computer networking under these conditions had the thrust to share resources (CPUs, databases), to provide distributed computer power to reduce communications cost, to aid in more effective use of companies' current computing assets, and to provide network users with more performance at lower cost.

The first true use of the collection of technologies for networking came along with ARPAnet. ARPAnet (Advanced Research Projects Agency Network) came online in 1969. It was built by Bolt, Beranek and Newman (BB&N) under contract to ARPA. The initial configuration consisted of four (4) nodes, located at the University of California at Santa Barbara, SRI, University of California Los Angeles, and at Utah. The system has grown now to include hundreds of sites and subnets connecting locations around the world into a large, resource-sharing network. Many users today take this network for granted when they send messages daily to colleagues throughout the system.

The ARPAnet provided a vehicle upon which much of the pioneering work for networks was performed. Network topology in terms of establishing connectivity between nodes and assigning link capacities was initially studied [Sauer, Kobayashi, Swartz], and network protocols for establishing, controlling, and releasing resources for interprocess communications were developed. Network flow-control algorithms were developed to control the volume of messages in the system. As a corollary network, routing algorithms were developed to aid in the selection of optimal paths for messages to traverse from source to sink nodes. Performance analysis methodologies and techniques were developed to aid in message delay and throughput analysis.

Communications in ARPAnet occurred by splitting messages into chunks called packets and then routing them one at a time (store and forwarding) from the source node to the destination node via a number of intermediary nodes.

These early networks were developed (1) to provide sharing of distant resources such as information repositories or processors within the constraints of cost and reliability of transmission links; (2) to provide high reliability by having alternate sources of computation, to maintain interprocess communication between network users and processors, to provide distribution of processing functions, to reduce transmission costs; (3) to furnish centralized control for a geographically distributed enterprise, providing for centralized management of distributed assets, and a vehicle to integrate dissimilar equipment and software, and to provide users with maximum performance at a minimum cost; and (4) to facilitate communications among users separate in space and time.

WHY LOCAL AREA NETWORKS?

The transition from the early resource-sharing networks to the local area networks was brought on by conditions such as researcher's vision to provide higher-capability architectures and user's needs for more services. The latter appears to be the more valid and the driving force behind local area network development. User's considerations for wanting a LAN include

- Combining resources
- Sharing information
- Linking remote sites
- Unifying databases
- Centralization
- Decentralization

To this point we have addressed large resource networks, their history, and their goals and uses. Local area networks have been mentioned, although we do not as yet have a good definition of what a local area network is. Many differing definitions have arisen over the years. Clark [1978] defines local area networks in relation to their components (simple network interface units), geographic separation (less than 10 kilometers), and their scope of services. Possibly a more precise definition or clarification deals with the wiring configuration, communication mechanisms, data rates, and environmental uses. Local area networks typically utilize wiring or communications media that are shared by a number of devices in the network (although there are some exceptions). They communicate in a peer-to-peer fashion. That is, devices communicate directly to each other, not in a store-and-forward fashion as in ARPAnet. The data rate in a local area network is much greater than in the traditional wide area networks.

Wide area networks data rates were in the range of tens of thousands of bits per second, whereas in local area networks rates typically are in the range of one megabit per second to hundreds of megabits per second. Local area networks typically employ some form of overall operating system environment. Whether it is a network operating system (put over local operating systems), a distributed operating system (one integrated over all sites), or a simple communications network server protocol, it exists on all nodes. Finally, local area networks seem to be employed in more integrated environments and problems. They are used to link office equipment together; as a controlling environment for a manufacturing plant; and as a corporate database backbone.In many cases, they form part of a high-speed, integrated computing environment, not simply as a means to send blocks of data, as in early nets.

LOCAL AREA NETWORK CONSIDERATIONS

When we as potential LAN owners look at acquiring a LAN, a few major considerations are needed to make a sound judgment. These major considerations are geographic, economic, social, and technological. Each has its own set of conditions to be looked at, but before any decisions are made, the potential LAN owner must make a determination that a LAN is needed. Why is the technology being considered, and what are the driving factors for going in that direction? Typically, one looks at LANs and their technology to solve the problem of sharing resources, making our present system more available, or uniting our database for greater overall efficiency and accuracy. These all map into one of the previously mentioned categories, as will be seen.

Geographic

One of the major contributing factors in selecting a local area network is to link up or connect a company's dispersed set of assets. No other technology provides an easier or more cost-effective means to unite a set of disjointed equipment into a "system." For example, if company A has 10 buildings, each with its own central computer, and a typical function of these machines is to use information from the various buildings' jobs to perform some computational task, then a LAN can provide a big win to the company. The LAN can be used to tie the 10 buildings' computers into a resource network and, with the proper software, a true distributed system. Thus, more computing power is made available, and integrated real-time data management is possible. This definitely beats the older method of performing data exchanges by moving tapes from one building to the other.

Another geographic consideration is to link up a set of PCs in a building. No matter what company you work for today, you will find personal computers. Each of these machines brings computing power to the end user, but again only to that user. With a LAN, all the PCs can be united into a large network providing the company the means to have a potentially paperless office (again, only if the right software exists).

Economic reasons stem from the added capabilities that come with a local area network. The LAN provides a means to offload work from overburdened systems to others on the network that are much more lightly loaded. It provides added services so that the multiple machines can integrate their disjointed databases into an integrated distributed database. Using such a capability, an enterprise can now make corporate decisions based on total corporate knowledge versus just a subset, and in today's age of information, the company that possesses the most up-to-date information wins. Beyond this, LANs provide a means to centralize control of computing assets by providing networkwide monitoring services. Many other economic reasons exist for purchasing a LAN, and some will be seen in later chapters.

Social

Social considerations deal with customer perception of worth. If your competitor has a LAN and uses it to build business and provide better services, then to compete you too must provide a LAN along with better service. Whether you view it as important or not, the customers' perceptions will ultimately drive you to the decision. An automatic teller machine that is linked to a bank's multiple databases, providing ready access to banking services, is one example of how advanced communications technology has affected us all.

Technological

The last consideration, though not one of the least, is technological. LANs provide services, and they do it in certain ways. The technical policies of operation of one LAN may not match a company's operations, and therefore could be a bad fit. The media used, if already in place in a facility, can drastically reduce the cost of a LAN installation. The technology employed in the interface units and other components can greatly affect the price and the performance. For example, if you are an office manager and you wish to link up your office PCs mostly for electronic mail and file transfers, then a cheap LAN such as an ethernet could be the best choice. On the other hand, if you are a designer of a military platform such as an aircraft carrier or submarine, where reliability and survivability are of paramount importance, then a customized LAN made of EMP (electromagnetic-pulse) hardened devices and with noise-immune fiber optics may be a better choice.

The major factors over all these considerations are the user's end needs. They drive the topology to consider the sophistication of the interface units, the volume and quality of LAN control, and applications software needed.

HARDWARE AND PROTOCOLS

Local area networks, strictly in terms of hardware, are composed of four major categories of components:

1. Topology (media and structure)
2. Interface units
3. Host computers
4. Servers and peripherals

The way these components are combined and the diversity of components within the classes are as numerous as the hundreds of companies building them. Most distinctions between the various local area networks focus on the media used, the topology and protocol, and the complexity of the interface units.

Media and Topology

Typical of most local area networks in use today is the use of serial data transmission media and techniques. This is due to the LANs' use in high-speed transmission, over moderately long distances, and the desire to keep wiring costs at reasonable levels. Due to this use, sophisticated line drivers and receivers (next section) have to be used, with only one driver/receiver interface unit per device. Transmission of the sequence of bits is typically strictly done digitally, using baseband or broadband signaling techniques. Baseband signaling is a technique in which the digital signal is transmitted in its original form with no changes induced by modulation techniques. Broadband signaling is a technique that provides a means to break up the transmission media into multiple channels over the same conduit by frequency division of the bandwidth [Fortier 1988]. Either of these techniques can be hosted on coaxial cable. Coaxial cable is a wire media made up of a central copper wire core encased in an insulator that is covered with a woven copper outer mesh.

Another favorite medium that utilizes digital signaling is twisted pair cable. This simple medium has been a big favorite for quite some time due to its simplicity and cost-effectiveness as a transmitter. A "Telecommunications Products and Review" article in April 1986 indicated that of 192 companies surveyed, 54% were using twisted pair; 20% baseband coaxial; 19% broadband coaxial; and 7% fiber optics. The major reason for this appears to be cost. Twisted pair cable is the cheapest, followed by coaxial cable and then by fiber-optic media. Fiber-optic media use light as the transmitting signal. The digital signal is sent as pulses of light: on is 1, off is 0. Fiber optics are excellent media for LANs because of their high bandwidths and immunity to noise. The problem with a fiber optics system is that it cannot be passively tapped as the wire bus versions can. Fiber buses need a passive star to which all nodes connect. Signals come into the star and are reflected out to all the other links.

In all the above cases we are talking about shared media that require control for correct use. This control is partially dependent on the topology used; e.g., the ring topology works best with token passing (sequential control). LAN control is typified by three major classes: sequential, contention, and reservation. Sequential-based control schemes operate by a sequencing of control for the media (access to the media) across the number of units in a strict physical/logical sequence. For example, if we had five units linked together in a string, then control of access to the media would rotate from unit 1 to 2 to 3 to 4 and to 5, then back to unit 1 and continue ad infinitum in this mode. That is, control passes from one site to the other, visiting each one once before reinitiating at the origin. This sequencing from site to site in a cyclic nature has been referred to as control cycle or scan cycle. A typical application of this control scheme is the token-ring bus communications control protocol.

The second class of control protocols is the contention-based protocol. This scheme uses active confrontation as the means to determine who has control of the media. One of the most well-known examples of this control mechanism is the carrier sense multiple access (CSMA) scheme. This protocol operates on global shared-bus media and performs control by collision detection. The basic mode of operation is as follows:

- All listen to media.
- All units that wish to send on idle media do so.

- They listen; if there is no collision (garbling of signal) during their whole transmission, then the message was successfully transmitted.
- If a collision occurs (what is sent is not what is heard), a few options are available. In CSMA collision-detection protocol, all STOP sending, wait some time, then try to send again.

The best-known implementation of this control protocol is found in the Ethernet system. It implements the simple CSMA collision-detection scheme with a binary backoff algorithm to protect against multiply recurring collisions. Details of this control scheme and other contention-based schemes can be found in Fortier [1988] and its references.

The third form or classification of control protocol is referred to as the reservation scheme. This class of control protocol is typified by a partitioning of the media bandwidth or time-scan sequence into components and the allocation of these pieces to the various devices based on user-predefined specifications. One major example of this type of control protocol was described in Jensen [1978]. The HXDP system (Honeywell Experimental Distributed System) was developed as the global bus structure with some number of devices attached to it. Control of the media was effected by the individual device's interrogation of their internal set of reservation slots. HXDP had 256 slots per total scan cycle, and the N available nodes were allocated anywhere from 0 to 256 slots of this, based on the system architect's view of node communications needs. Internally, the control occurred by the sending of an update signal that caused each node to increment its control scan pointer to the next slot in its scan memory. If it has a stored one in this location, it gets control of the media; otherwise, it waits until the next control update. At any time, it must be guaranteed by the initial allocation that only one device gets control of the media at any time. This class of control protocol allows for the tuning of the network to user needs, but possibly at a cost of wasted transmission time for the times when the node given control has nothing to send.

With these three basic control styles, we can describe a wide variety of control schemes that provide services from simple linkage to highly resilient, real-time service. For example, we can classify and define within this scheme, the basic pure aloha collision-transfer scheme, or the more resilient CSMA/CD p-persistent, and persistent schemes. Additionally, more robust schemes such as the DSDB_DOT_OR [Fortier 84] can be described, as can the token bus, token ring, and others. Detailed control protocol discussions were provided in Fortier [1988] and Stalling [1986], and therefore will not be discussed at this point.

As indicated previously, a tightly related topic to control is media topology (interconnectivity). Topology refers to the structure by which the various devices are connected together and the view of this interconnectivity. The typical structures are the ring, bus, star, mesh, and hybrid topologies.

The ring topology is best described as a set of interface units with a single input and output that connect to their nearest neighbor's output and input, respectively, and end with the initial unit's input connected to the final unit's output (Figure 1-5).

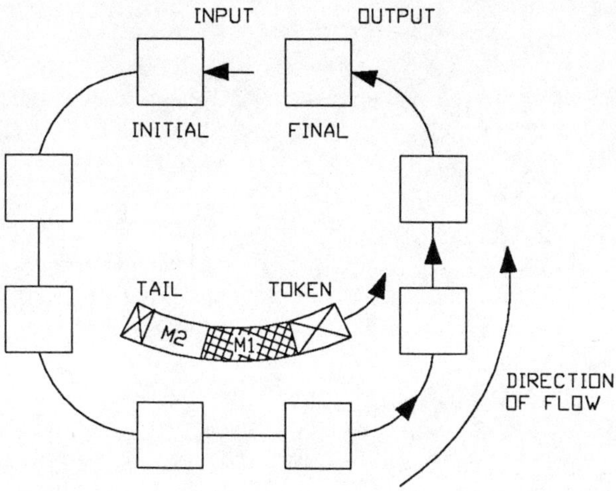

Figure 1-5. Ring Topology.

Data and control flow over the same media conduits and in a single direction only. Control in such a topology is typically performed by a token that flows from unit to unit. As the token arrives on my unit, I am allowed to place a message onto the ring, typically on the tail of the present message stream (Figure 1-5). Variations of this scheme include a dual rotating ring (DDLCN), the insertion ring, and slotted ring control schemes. The dual ring allows us to have higher availability of access to the media. At worst, we would have to wait half of a rotation to get to the media versus one full rotation. The problem comes in determining the access scheme to the media by the interface units and their ability to determine which of the media channels to use at any one time. The insertion ring still utilizes the rotation of the ring message trains as control boundaries, but provides a means to acquire use of the media in a more random scheme. As a message begins to enter my node, I delay it, insert my message into the train, then ship the original message out after mine. This scheme allows the nodes to get higher utilization of the media. A similar scheme for increasing utilization is the slotted scheme. In this control protocol for the ring topology, we break up the ring into slots (Figure 1-6).

These slots provide the places in which we can place our messages. As the slots come across my interface unit, I examine the header field; if it is marked full, I cannot use this slot to send a message. If it is marked empty, I can. Once I see an empty slot, I mark it full and ship out my message. When the message comes back around, I remove it and mark the slot empty once again, making it available for others to use. The problem is that if the recipient is one node over in the sequence, the slot is still held for a full rotation cycle.

A second topology is the bus topology, often referred to as the global bus topology due to the nature of its operations (shared media). Bus topologies typically implement a contention-based control scheme even though they have been constructed with all

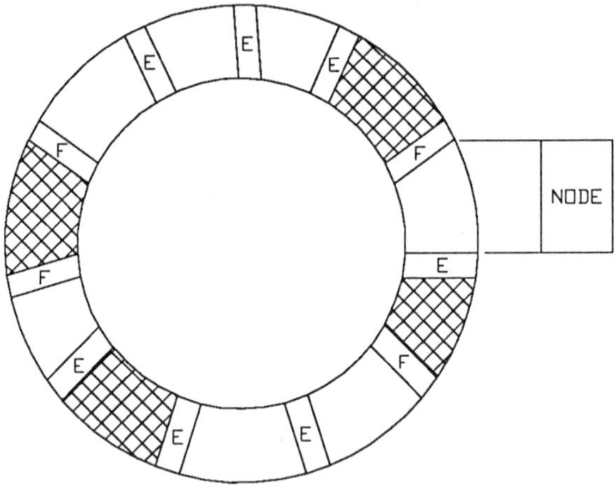

Figure 1-6. Slotted Ring.

three control mechanisms. The simplest form of the global bus topology is seen in Figure 1-7. This topology consists of a single media link with devices strung from it using "T" interconnects. The media are a broadcast type and require the ability to be tapped into easily. Typical media used for this topology are twisted pair and coaxial cable. Fiber-optic cable can be used, but it must be strung as a passive star with all nodes cabled to it. The global bus topology has been employed in some of the most widely used local area networks; for example, the Ethernet system utilizes a global bus topology, as well as many others. This topology also offers the best opportunity to support a greater diversity of control protocols beyond Ethernet [IEEE Std 802.3] and other CSMA/CD schemes. The global bus has been used to implement a token-passing scheme [IEEE Std 802.4] and reservation schemes [Honeywell HXDP, Jensen 1978; DSDB Fortier 1984]. Beyond the ability to support a wide range of control techniques, the global bus topology lends itself to an easy installation. It can be routed and installed in much the same way a phone system would be, with a main trunk line through a building and subtrunks for floors (Figure 1-8).

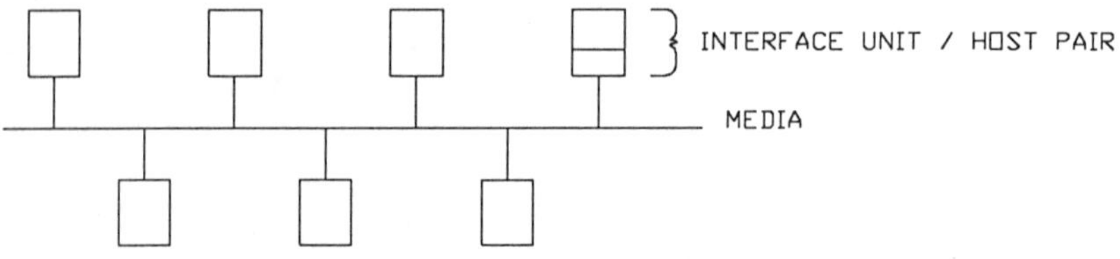

Figure 1-7. Global Bus Topology.

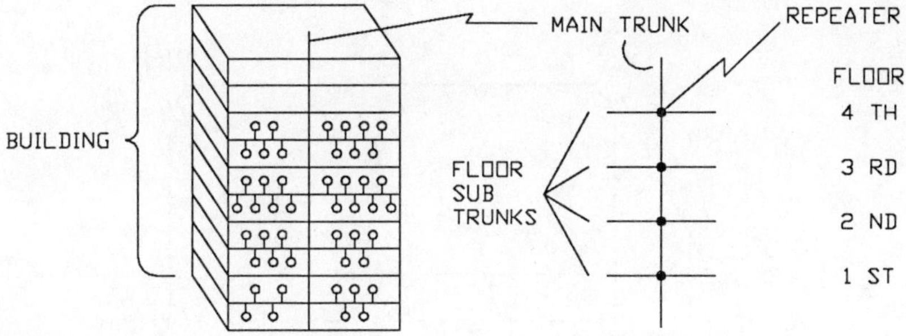

Figure 1-8. Global Bus Building LAN.

A third class of topology is the Star topology. In this interconnection scheme the node of the local area network is interconnected via an active hub device (Figure 1-9). All communications are routed through and are controlled by the active hub. This topology requires the LAN to have additional software for routing, flow control, and contention resolution, as well as more hardware associated with the linking media. Reliability is a concern in this topology, since the loss of the hub controller will cause loss of all communications, whereas in the other topologies, there are easy work-arounds dealing with fragmentation of the network. The problems with star topology have led to its demise as a viable interconnection scheme, and networks using this scheme tend to be experimental.

Stars may reappear as communications technology matures and VLSI can be applied to the hub controller solution.

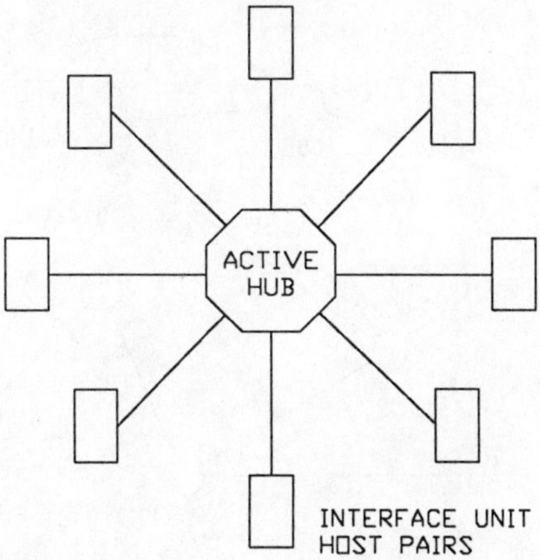

Figure 1-9. Star Topology.

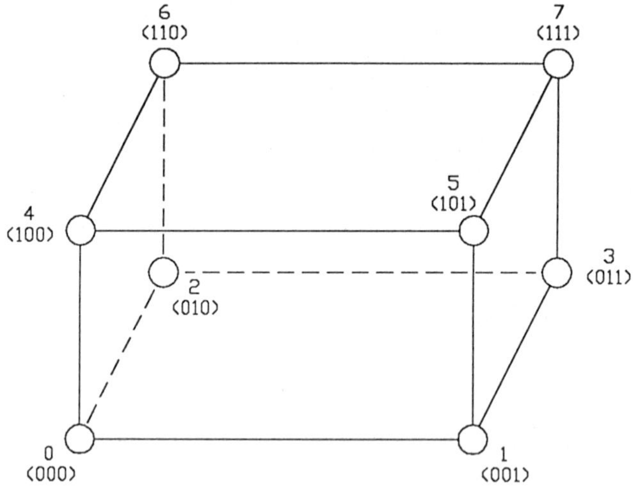

Figure 1-10. Hypercube Topology.

The fourth topology to be considered is the irregular or mesh topology. This topology is typified by some number of point-to-point connections, configured into a structure. For example, the hypercube (Figure 1-10) is a realization of a mesh type of topology. Another example, the random mesh, is shown in Figure 1-11. Topologies of this type use point-to-point interconnects and store-and-forward type protocols to communicate information from source to destination, as was discussed for the ARPA-net system.

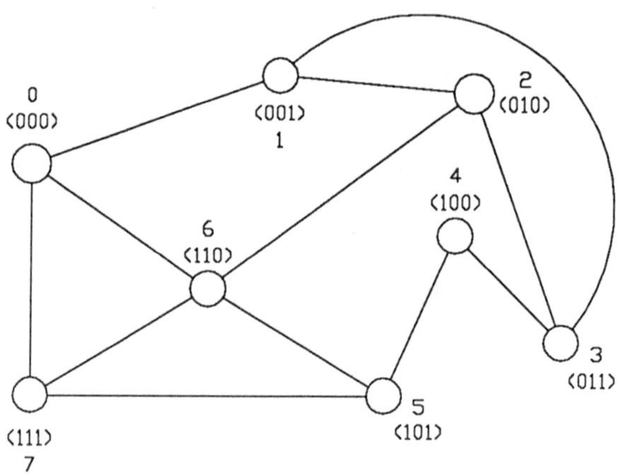

Figure 1-11. Random Mesh Topology.

For example, in Figure 1-10, if the source is 0, (000), and the destination is 111, then the message can be sent over one of many possible paths:

000	--->	001	--->	011	--->	111
000	--->	001	--->	101	--->	111
000	--->	010	--->	011	--->	111
000	--->	010	--->	110	--->	111
000	--->	100	--->	101	--->	111
000	--->	100	--->	110	--->	111

In all cases the distance the message must travel is three HOPS or node steps. The problem with this topology and all its flexibility is to determine which of the routes to use. What is the optimum (lowest cost in terms of delay)? This problem is referred to as routing. Routing is the process for selecting one of the possible paths for a message based on some optimization criteria or process. Typical mechanisms include hot-potato routing, flooding, delta, and many more. Details can be found in Fortier [1988] and Tannenbaum [1982]. An additional issue with this topology is that of flow control. This deals with controlling the flow of messages through the network to provide correct and consistent service. Flow control looks to limit the allowable traffic on a link or within the system. Techniques such as preallocation of buffers, windowing, and others have been used for the solution of this problem. Again, the previous references have greater details on these topics.

The last class of topology is a variant or combination of all of the above. The hybrid topology consists of components taken from the previous four basic topologies. For example, we could use a ring as a backbone network, with buses on the peripheries (Figure 1-12a), or a star with the buses (Figure 1-12b).

The diversity of interconnections within this class of topology is very large. The limit is in the eyes and imagination of the builder and implementers. This class of topology lends itself to tuning; that is, we can select the best features of each topology

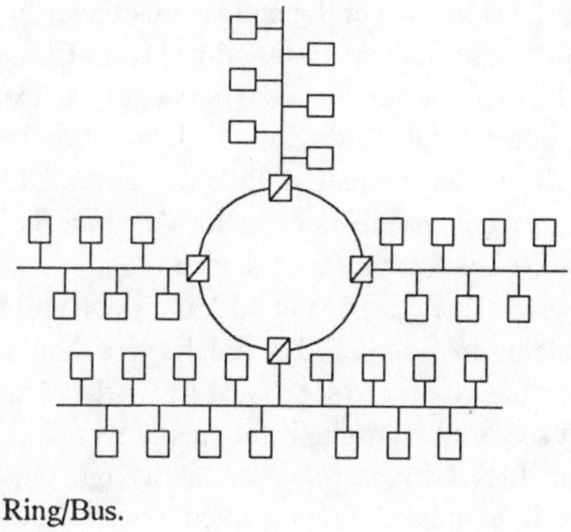

Figure 1-12a. Hybrid Ring/Bus.

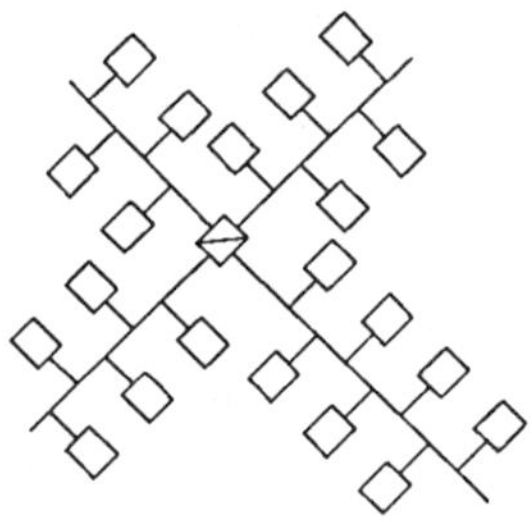

Figure 1-12b. Hybrid Star/Bus.

and incorporate them into one integrated topology. More will be said about this type of topology and its problems when we look at bridges and their software in later sections.

Interface Units

To connect a host device into communications media and make it into a network requires the use of an interface unit. A network interface unit (NIU) provides the mechanisms for the attached devices to send and receive information over the computer network in a predefined, controlled fashion. All interface units for local area networks possess three basic components (Figure 1-13):

- Network adapter
- Input/output processor
- Host adapter

The network adapter has the function of performing the low-level physical transmission and reception of signals over the media and converting these for internal use.

The input/output processor has many functions, from forming messages for transmission over the media to preparing received messages for host use. The IOP controls all interaction with other network devices and provides for queue management, message handling, priority determination, link allocation, addressing, error management, and transmission. The host adapter provides the interface and buffering for data transfers to and from the host computer or device.

A more detailed view of the generic network interface unit (Figure 1-14) indicates that the three components are composed of much more than this.

The network adapter has circuits to send and receive data (bits) over the media based on the design being used. For example, if the media were a fiber-optic link, then the circuits would consist of laser drivers to output bits as light pulses and a light-sensitive receiver to pick up the light pulses being received. For the Ethernet coaxial cable case, the hardware would be simple modulators-demodulators to send electrical signals over

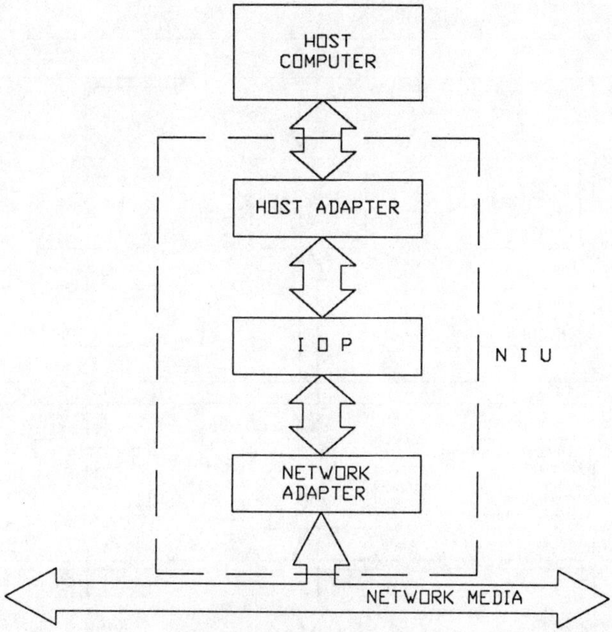

Figure 1-13. Generic Network Interface Unit.

the cable and be able to extract them from the cable. Bus interface hardware controls the routing of this data to or from the appropriate buffers set up by the controller for the outgoing message. It also adds any pertinent data (checksums, etc.) that are required by this implementation. The network adapter has buffers that are used to store data for reception or transmission; they are tuned to the particular network requirements. Additional hardware is used to provide DMA (direct memory access) services to and from the buffers. The network adapter controller provides for the low-level line-control protocols (address detection, etc.) and coordinates the action of the NA (network adapter) components.

The input/output processor component has the job of implementing (carrying out) the policies of media access and control included in this job and processing to perform message handling:

- Construction/reconstruction of messages
- Header determination
- Priority ordering of messages
- Allocation of media
- Sequential control
- Contention-based control
- Reservation-based control
- Addressing (source and sink)

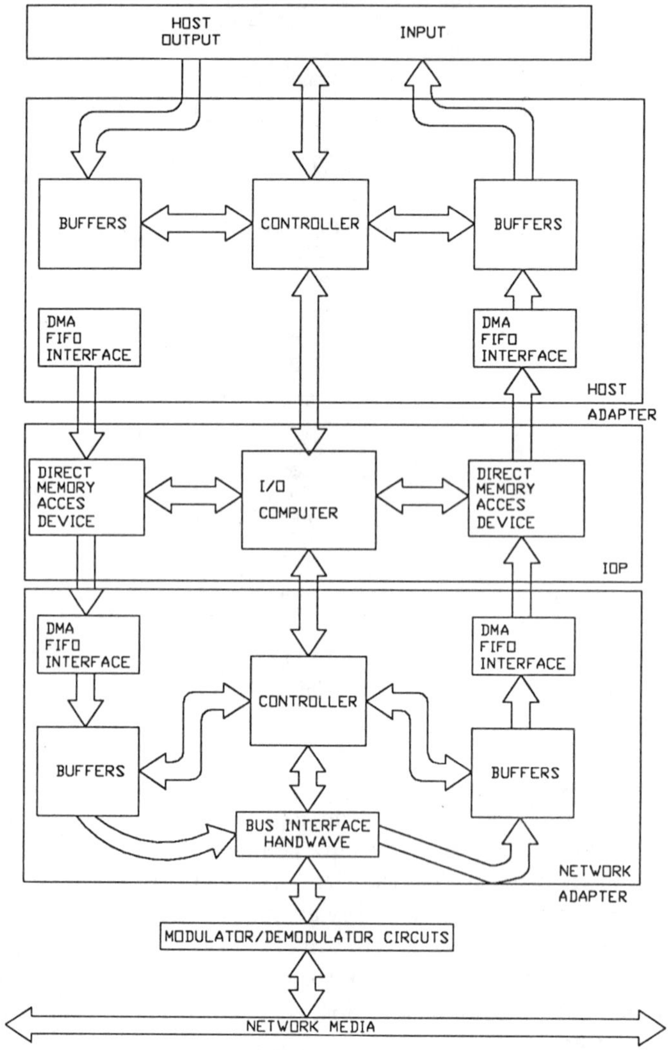

Figure 1-14. Detailed NIU Architecture.

- Error control
- Error management
- Routing
- Flow control
- Security access
- Queue management
- Buffer allocations

This component is the heart of the network interface unit and gives the network its character. That is, it makes the network behave as it was designed to.

The host adapter provides the interface and control to make the NIU (network interface unit) available for a wide variety of devices.

NIU Standards

The ISO/OSI reference model of network protocols led to the development of low-level hardware standards. These standards cover the two lower layers of ISO (next section) including the physical layer and data-link layer.

The three main products of the IEEE 802 Committee for Layer I are the 802.3, 802.4, and 802.5 standards. The 802.3 standard implements the carrier sense multiple access/collision-detection scheme of control, as described for Ethernet previously. The 802.4 is a token-passing bus access protocol that implements token-passing as described earlier, but on a global bus topology. The manufacturing automation protocol (MAP) being pushed by General Motors is a leading contender to use this protocol and standard. The 802.5 protocol specifies a token-passing ring access method as described earlier, with IBM's implementation of this protocol being set as the de facto standard for this architecture. Table 1-1 illustrates implementations for these protocols by the manufacturers of integrated circuits. This sample is but a touch on LAN interface units. Fortier [1988] lists a large volume of LAN vendors and their products in this area.

HOST DEVICES

Computer networks are built to support the interconnection and communication among various devices in an enterprise. These devices come in a variety of shapes, sizes, and functionalities. The network must be designed so that this diversity of devices can be linked together into a functioning computer system. The LAN software in conjunction with its hardware provide the glue that allows us to have microprocessors mixed with supercomputers on the same network. This type of service implies a heterogeneous environment, allowing nonsimilar devices to be supported—it causes extra overhead and complexity on the LAN, but benefits users by simplifying their use of the network. The user simply plugs in and goes with no, or very little, modification to their assets. On the other hand, some networks such as company proprietary ones support a homogeneous environment. This implies that the network may only support this company's machines and nothing else. Such networks can simplify their interfacing hardware and software, but at a cost of eliminating the ability to connect any other vendor's devices to their network—which could be a big limitation down the road when new resources are needed.

Table 1-1. IEEE-802 LAN ICs

Manufacturer	Part	Function	IEEE Application
Advanced Micro Devices	7996	Transceiver	802.3
	7990	Controller	802.3
	7992B	Serial Interface Adapter	802.3
EXAR	XRT82515	Transceiver	802.3
	XRT82C516	Encoder	802.3
Intel	82586	Coprocessor	802.3
	82501	Encoder	802.3
	82588	Controller	802.3
Motorola	MC68184	Broadband Interface Controller	802.4
	MC68824	Token-Passing Bus Controller	802.4
National Semiconductor	DP8390	Network Interface Controller	802.3
	DP8391	Serial Network Interface	802.3
	DP8392	Coaxial Transceiver	802.3
SEEQ Technology	8003	Data-Link Controller	802.3
	8023	Manchester Code Converter	802.3
Signetics	NE5080	FSK Modem Transmitter	802.4
	NE5081	FSK Model Receiver	802.4
Texas Instruments	TMS38030	System Interface	802.5
	TMS38010	16-Bit Communication CPU	802.5
	TMS38020	Protocol Handler	802.5
	TMS38C51	Ring Interface Controller	802.5
	TMS38052	Ring Interface Transceiver	802.5
Thomson-Mostek	MK68590	Controller	802.3
	MK68591	Serial Interface Adapter	802.3

Table 1-1. *Continued*

Manufacturer	Part	Function	IEEE Application
	MK68592	Serial Interface Adapter	802.3
	MK5030	Controller	802.3
	MK5032	Controller	802.3
	MK5033	Manchester Encoder	802.3
	MK5034	Manchester Encoder	802.3
	MK5035	Manchester Encoder	802.3
	MK5036	Manchester Encoder	802.3
Western Digital	WD83C503	Controller	802.3
	WD83C510	Controller	802.3

Devices connected to local area networks run the gamut from simple I/O devices, such as printers and FAX machines, up to large supercomputers and database machines. Each has its own unique interface requirements and levy different requirements on the LAN.

Supercomputers

Supercomputers are a class of computer system that provides computing power in the range of hundreds of megaflops (million floating-point operations per second). These machines are used for processing intensive applications such as weather prediction, large simulations, aircraft design analysis, spacecraft analysis, astronomy and medical applications, and numerous others. Most machines of this class are highly parallel machines that do numerous computations concurrently. They require much setup time to perform their task in terms of loading memory, registers, etc. Examples of such machines are the Cray I and II computers. The Cray I has shown that it can perform computations at rates exceeding 100 Mflops in particular scenarios. The Cray consists of 12 tightly coupled functional units; is pipelined; has specialized vector processing units; provides linked data paths from output of one unit to input to another, thereby avoiding primary memory accesses; and has an extremely fast computing cycle of 12.5 ns. The Cray II is an improvement on the Cray I and has been touted as being more than twice as fast as the Cray I. These machines put special constraints on a LAN to be able to provide high-speed I/O and large buffering for loading and unloading the machine for the network user's specialized computing.

Mainframes

This class of computing device is representative of a general purpose computer of power less than the supercomputer and greater than most minicomputers. Representative of this class of machines are the IBM 4300 system series, the IBM 3070 line, the CDC 6600, and Digital Equipment Corporation's 8600 series of computers. These machines are designed as large, general purpose data crunchers. Mainframes are typified by wide data paths, usually greater than 32 bits, frequently 64 bits or more, large memories (millions of words), and much secondary storage (large banks of disk drives, tape units, and drive units). The mainframe machines more closely relate to the supercomputers in the sense that they typically use some of the same tricks to speed up; for example, large banks of cache memory, special registers, multiple processors, and specialized computation devices can often be found in them. As in the supercomputer case, extra treatment will be required from the network to support their high data-rate requirements.

Minicomputers

The term minicomputer can be quite misleading because the term covers a wide spectrum of midrange-performing machines. Minicomputers typically are medium priced machines ($20,000–$200,000) that provide general purpose computing services with moderate speed of 10 Mflops or greater, that come with large memories of four megawords or greater, and that support a wide range of I/O devices. They typically support word sizes of 16–32 bits in length. Minicomputers provide lower-cost computing power for small- to medium-sized business applications.

Workstations

The workstation is a relatively new phenomenon. Workstations typically are high-range personal computers with 32-bit microprocessors as their basic processing engine. Additionally, they come with large memories—one to four megabytes—with extensive hardware support for graphics, plotters, and many other integrated peripherals. The clock rate on most of these devices is > 20 MHz and they come with extensive software applications, such as CAD tools and graphics packages. The most distinguishing characteristic of the workstation is its integration with a network environment. Representative workstation vendors include Sun Microsystems, Digital Equipment Corp., and Apollo Computers.

Personal Computers and Microprocessors

The personal computer (PC) is the latest of the network devices. PCs find their way into all walks of life. They are used to produce books (such as this one), keep accounting accurate, as a planning tool, and for computer-aided design. They are used in small-business retail sales as well as in large conglomerates as individual productivity enhancement devices.

PCs tend to be of low- to mid-power with clock speeds from 8 to 20 MHz; they typically come with greater than 640K bytes of memory and with a single floppy drive or up to numerous hard disks. They are single-user systems, and in large businesses are typically linked into resource nets. The typical use of a network in this type of system is for file transfers; however, this author feels as time goes on we will see these networks becoming large, very powerful, fully integrated, distributed computing systems.

Microprocessors, the basic component of all PCs, have become an integral part of everyday life. We find them as controllers in our microwaves, VCRs, televisions, cars, phones, and many other appliances that we use on a daily basis. These, too, will be linked into networks to provide house control and data management.

Beyond the basic computers, the local area network must provide services to host device servers such as printer banks, database machines, file servers, data collection devices, monitors, archival systems, and numerous others.

REFERENCE MODELS

As the networks themselves came into being, so did efforts toward standardization. Committees were formed with the charter to define a set of standards upon which networks would be specified and built, and around which many vendors could develop products. The International Standards Organization (ISO) formed a Subcommittee for Open Systems Interconnection (OSI) with such a charter. The committee set upon the task of defining a reference model that described the functions of a standard network in abstract terms. This model consisted of seven layers (Figure 1-15) with each layer specified to perform various functions, as follows:

Layer 1: The Physical Layer

- Transmission of raw bits of data over a communications channel.
- Design issues that deal with the mechanical, electrical, and procedural interface to the subnet.
- Protocol examples RS-232, X.21, X.25, 802.3, 802.4, 802.5, and 802.6.

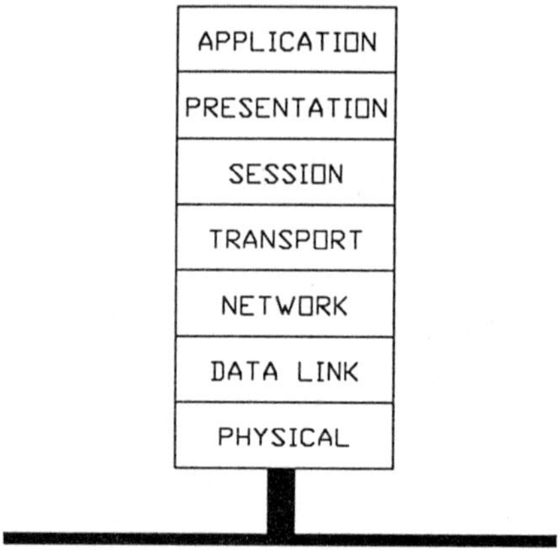

Figure 1-15. ISO Reference Model.

Layer 2: The Data-link Layer

- Uses the physical layer to develop an error-free transmission line for the network layer.
- Data framing, sequencing, and acknowledgement.
- Error handling (error detection and correction, example 802.2).

Layer 3: The Network Layer

- Controls the operation of the subnet.
- Packet routing/flow control.
- Breaks a message into packets (IEEE 802.1).

Layer 4: The Transport Layer

- Performs end-to-end or source-to-source protocols.
- Assembles and disassembles messages from lower and upper layers.
- Creates a distinct network connection for each message.

Layer 5: The Session Layer

- Provides the main user interface to the network.
- Provides two classes of functions:

  1. Binding and unbinding between two processes; e.g., session initiation and termination.
  2. Control of data exchange, delimiting and synchronization of operation (session dialogue services).

Layer 6: The Presentation Layer

- Provides services for the application layer to interpret the meaning of data exchanged. This includes management of entry, exchange, display, and control of structural data, (e.g., encryption/ decryption)

Layer 7: The Application Layer

- Concerned with the problems of network application.
- Network transparency; problem partitioning; distributed databases.

To better illustrate how such a layering could be used and to understand what these layers imply, I will describe an ISO example of sending a letter; in this case I am sending a resumé for a new job.

The Application

The network application is to transmit a job-seeking letter to Company X with my resumé. The presentation layer provides services for the application layer to interpret or enter network data in any form. Since my secretary reads all incoming and outgoing mail, I write in French (Figure 1-16):

"Mon Ami . . ."

The session layer is the user interface to the network; it provides initiation/termination of session and control of the data exchange. In this example, I know the mailman comes at 10 a.m., so I wait for him at the pickup point to hand-deliver my application letter.

The transport layer takes over. This provides the end-to-end protocol that creates a distinct connection for each message. My letter consists of a five-digit zip code that the post office increases to nine for greater granularity.

The network level controls network operations such as breaking messages into sendable pieces. In the example, the post office uses carrier pigeons that can only carry 1/2 oz. of mail at a time. Since my resumé and application are more than that, actually 1 oz., they cut it in two and ship it with two birds. The birds have a predefined route that keeps my resumé from unwanted hands (i.e., my bosses). The birds follow this route to protect themselves and get rest, food, and water.

The data-link level provides for error-free transmission. To provide this service, the post office uses another pigeon to send a message back that the letter got there all right. Additionally, to provide this service, the postmaster keeps a copy in case the pigeons get hit by lightning or meet some other terrible fate en route.

The physical level provides the raw transmission services. In the example, this service is the little pigeons flying through the air with my letter.

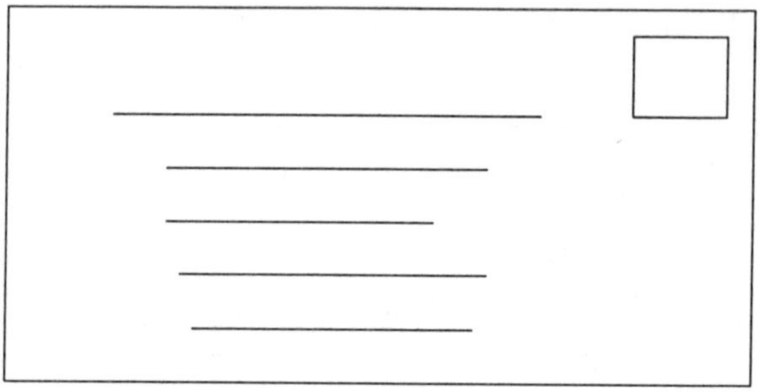

Figure 1-16. French Letter for Job at Company X.

In reverse order, once the birds arrive at the receiving site (the other postmaster's station) where Company X is located, the data link (the postmaster) dispatches the "ack" pigeon, the network puts the two pieces together, the transport checks that this is the correct zip code, the session (mailman) delivers the letter to Company X at 3 p.m., the presentation (Company X's interpreter) translates my French, and the application (personnel) determines I'm not qualified.

This simple example illustrates how applications use the OSI model to effectively communicate over a network. The details of the layers and their functions are provided in Chapter 6 and references.

Book Model

Since the inception of the ISO/OSI standards, many variations on this standard have arisen: the SNA (Systems Network Architecture) of IBM; the DNA (Digital Network Architecture) of Digital Equipment Corporation; the IEEE 802 Committee interpretation of ISO/OSI; and, yes, I too have a pseudo-standard that is derived from or extracts concepts from these.

Most systems that I have dealt with use the concept of layers, although the distinction of what lies in each is different from the OSI standard.

To categorize the book's structure, I use the following five-layer model (Figure 1-17), more closely related to LAN communications:

1. Applications
2. Systems Management
3. Transport/Monitor
4. Data Link/ Network
5. Physical

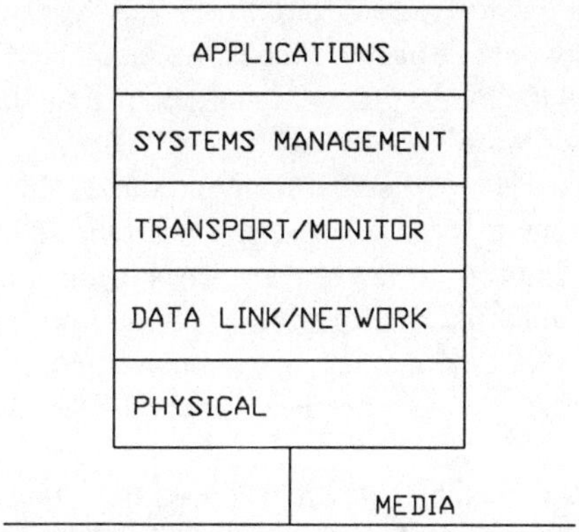

Figure 1-17. Book Five-Layer Model.

The physical layer provides the mechanical and electrical characteristics for interfacing onto a media, the timing associated with communications, and the definition of control within the media.

The data-link/network layer provides services necessary to address units and so processes objects on the network to control routing, flow control, and low-level errors of these units of transmission.

The transport/monitoring layer provides low-level systems management of communications assets, configuration and reconfiguration of the media, as well as performance monitoring and fault isolation services.

The systems management layer provides the view of what the network looks like to users. That is, it provides for the Operating System services, data management, security, and information exchange services. This layer is what the user applications need to know about the network to make it work.

The top layer as in the OSI model is the applications layer. This layer is where user applications reside as well as network-provided applications. The remainder of this book will address these layers in greater detail, ending with an examination of some presently available and proposed software for local area networks.

LOCAL AREA NETWORKS SOFTWARE

A LAN should provide integrated services and applications over a wide range of devices with no added user interaction. Users want to sit down at a terminal on a LAN and use resources (programs, hardware, information) on micros, minis, mainframes, supercomputers, and special purpose devices without worrying about where they are or how the communications and management are established, coordinated, or performed and without having to learn a different user interface and environment for each of these systems. This would be heaven, and in the future it will be possible, although now we are still forced to pay the price for nonhomogeneity in resources, LANs, and the attendant service software's overhead.

Present LAN software is composed of many products, although each is individual in terms of its requirements, performance, and mapping to networks. But all is not bleak; there are beginnings of this heaven on a LAN forthcoming. Software packages have been around that unify multiple operating systems and their networks, providing for communications among nonhomogeneous environments. As time goes on, the sophistication of this software will increase to meet the needs and desires of the user community.

LAN software is the media that will achieve this transformation, and in the mean-time it provides us with the functionality exhibited by LANs. As LAN software matures, it will provide users with the features and services they need in order to concentrate on the work at hand instead of on the network in between.

LAN hardware has been stabilizing, leading the march for software vendors to come in and open new doors. LAN software is becoming the computer market's new frontier for product innovation and growth. The future will see a rich, new sea of products that will operate on a variety of machines and be invisible to users in terms of a network as its core. All software will run on your native, known operating system, and even if it is running remotely, it will look and feel as if it is running on your local machine.

This rise in software development is a response to the pressure brought to bear on LAN vendors to provide higher-powered capabilities to users of these networks. The LAN software developed will be the heart of the system and will provide its performance. This performance can come in one of four ways via a network and software:

1. Integrate present stand-alone devices
   a. Provide for data transfer only.
   b. All will need copies of software they will run.
2. Provide remote access to software on other machines.
   a. Can be shared.
   b. Find-and-go type service.
3. Provide local integrating access to all software.
   a. Requires global operating system to hide all intricacies of network and remote software access/operation.
4. Provide all of the above.
   a. A little of each.

Each of these choices of LAN environment has an implication on the software that can run. It may not be possible to take your typical PC single-user software product into a network environment and have it work properly. Complex safeguards must be written into the software to maintain orderly access and operation of it. Typically, off-the-shelf, single-user software will not operate in a LAN environment unless access and control software are provided to service multiple requests and provide for controlled use. If this were not done, then multiple network users trying to access a single software entity will typically cause it to crash. If it doesn't crash, then some erroneous computations will inevitably result, leaving both users worse off with the LAN than without it. The LAN software must enhance the user's productivity by making all data and programs available for shared use. If not, then, as LAN software designers, we have failed. The LAN software must provide integrated applications software providing compatibility between network applications and local applications software. Integrated LAN applications software designed for the multiuser environment offers the best opportunity for true office automation and distributed processing at the lowest cost per user, and provides the best return on investment for the LAN and its user base.

To provide such services the LAN must possess a variety of software components that perform various layered functions. LAN software exists from the lowest physical level up to the user applications software and system applications software. LAN software provides varying service to physical, data-link/network, transport/monitor, systems management, and applications layer components. Without these underlying services many of the upper-layer components could not do an effective job.

LAN software is most conveniently thought of as layers, like stories of a building. The layers are connected via well-defined interfaces designed so that the lowest layers can be replaced without redesigning the upper layers. Users of networked software applications deal with the highest layers, which are those least likely to change as network hardware technologies evolve.

At the physical level, there is not much in the line of software found in any implementations and therefore we will be ignoring software here.

The data-link/network layer provides basic services for all upper-layer software. Its main function is to supply communications management, interprocess communications services, and routing and flow control for the upper-layer users. Included in this level under communications management would be software to detect and log errors, monitor hardware status, find locations, establish connections, determine routes, allocate/deallocate buffers, assign channels, initiate read/write routines, coordinate frames to/from low-level communications memories, detect/recognize traffic from node to node, and formulation of messages for transmission. IPC (interprocess communications) mechanisms handle the low-level interface with the hardware and provide the protocols for end-to-end communications of the data bits.

The transport/monitor level performs systems management functions that handle the detection and isolation of faults in the system, assess availability of systems components for use (hardware or software), provide maintenance of a map of the configuration of operational hardware and software, manage workarounds necessary on failures, and maintain a high state of availability to keep the system software running.

The system management layer is where an integrated computing system is built from the basic communications and support software mechanisms. Synchronization primitives, basic remote invocation primitives, data management, operating systems, services, and interprocess communications primitives are provided. These elements work together to provide user and system applications the insulation from the network and the basic client-server interfaces to perform computations as if operating on a monolithic computing system.

The applications layer provides the area where system applications as well as user applications are performed. This layer acts the same as a traditional PC or other computer-user applications environment where one can use programs, write programs, and do a variety of other useful things. For example, programming languages should be provided that allow for the construction of algorithms that will run on multiple machines; compilers, linkers, and loaders that do the translation and mapping must also be provided.

Each of these layers provides services to the others upon which they build their own services. With the proper mix of LAN software, large pools of computers and workers can be linked together and work on single, large computational tasks. For example, with the proper applications program we could link an entire manufacturing company together, providing for user access and control of all aspects of data collection, real-time computer control, data processing, and myriad other computer-based operations. Designers could share their designs by electronic media versus the traditional paper. They would work more efficiently on large designs, since all the pieces could be available for their perusal, thereby aiding in interface designs and other feats requiring coordination. The LAN software provides the means for these users to share information and resources, no matter what the end hardware is (at least in the perfect case).

OVERVIEW

The previous section as well as this one provide some of the basic notions of what a LAN is and how LANs have been used and will be used in providing communications among a variety of devices. Additionally, the concept of LAN software was introduced, and a discussion of layering and what can be expected of software at the layers was presented. LAN software was introduced as software written to operate within a local area network, or software that has been modified by some means to operate within a network.

Chapter 2 introduces the applications layer and all of its generic software. In particular, that chapter looks at applications software from six categories:

1. Resource-sharing software
2. Model/simulate/forecast/planning software
3. Develop/maintain/execute software

4. Information management software
5. Design software
6. Control software

The coverage within these categories is on a generic basis, where each type of software will be discussed in terms of its operations and from a network user's viewpoint. The discussion is meant to provide the users with an understanding of the product categories' functions without going into specifics of any one product or technique.

Chapter 3 looks at systems management software. The goal of that chapter is to introduce the reader to the class of LAN software that provides for network management of assets and data. In particular, Chapter 3 addresses:

- Distributed operating systems
- Distributed data management systems
- Information exchange protocols
- Security

Again, as in the other chapters, the various categories will be discussed in terms of generic architectures and features. We will look at the general structure of a distributed operating system. The three basic models of design for these systems are used as user applications construction basics. Distributed data management will focus on the generic architecture of a distributed data manager and how those pieces can be built and supported with a LAN.

Chapter 4, Transport/Monitoring, deals with the detection of faults and determination of performance along with configuration management, availability management, and reconfiguration management. The main aspect of this management is to keep a mapping of resources, their roads and status, to be able to assess how to best reallocate resources in the event of a failure.

Chapter 5, Data Link/Network, looks at LAN software dedicated for the determination and management of routes in the LAN and between LANs. Covered in this chapter will be:

- Routing
- Flow control
- Addressing
- Bridges
- Gateways

Chapter 6 looks specifically at companies, their offerings, and how they come together.

2

APPLICATIONS

INTRODUCTION

The applications layer, the topmost in the seven-layer ISO/OSI systems interconnect model as well as the topmost in the book's five-layer model, is by far the hottest topic in the LAN industry. This layer is the final frontier, the one where users get full use of a network and within which they can construct the future distributed processing products. Simple products such as file transfer protocols, virtual terminal protocols, message handling systems, transaction processing, job transfer and manipulation, remote databases access, and remote procedure call mechanisms are coming into being—as well as more advanced products that are built on top of these services. The applications layer and the user applications applied to it provide the framework to build onto truly integrated environments.

USER DATA-PROCESSING FUNCTIONS

As was indicated above, users writing applications on top of the basic applications layer services can build a wide range of software. For example, with basic file transfer capabilities we can construct a crude distributed database management system. Or with remote job entry and call facilities, we can create truly distributed programs to do a variety of jobs. Remote synchronization and concurrency control lead to true distributed database systems.

User applications can be categorized into six major areas:

1. Resource sharing
2. Model/simulate/forecast
3. Develop/maintain/execute
4. Information
5. Design
6. Control

These rough categories provide a means to organize the wide range of user local area network software products. Resource-sharing software deals with one user using the assets of another in some shared/controlled fashion.

Model/simulate/forecast software categorizes LAN software developed to aid in the analysis of LANs or any other environment using distributed assets.

The "develop/maintain/execute" category captures the traditional mainframe software engineering environment consisting of compilers, linkers, loaders, debuggers, etc., with the big difference being that these; i.e., LAN-based development tools were constructed to use the LAN's multiple sites and communications capabilities to their advantage by design. For example, the compiler may use the LAN topology and node distribution to compile programs into distributed pieces to be executed concurrently, thereby enhancing user productivity. The linker aids in the location of distributed pieces of executable images and bringing them together.

The information category deals with LAN software that controls the interaction of data for some end product. Typical of this class of service would be memo mailers, bulletin boards, database servers, and other forms of information transfer. The design category deals with LAN software that provides computer-aided design and manufacturing, where the design is coordinated over a series of machines and linked via the LAN into a large CAD/CAM database and designer tool for multiple designers to utilize.

The final category of applications software is control software. This could be both automated and interactive, the function being to control some devices on the LAN from other points via software interactions. Examples such as robotics and manufacturing control come to mind immediately, although others do exist.

Resource Sharing

The basic component of any distributed computer system is its interprocess communications. Without an effective means to communicate, no true distributed processing or resource sharing could be effected. The three main classes of remote communications and invocation are the remote procedure call, message passing, and object invocation mechanisms. These are all typically built on top of operating systems interprocess communications facilities (IPC, Chapter 3). With any of these techniques

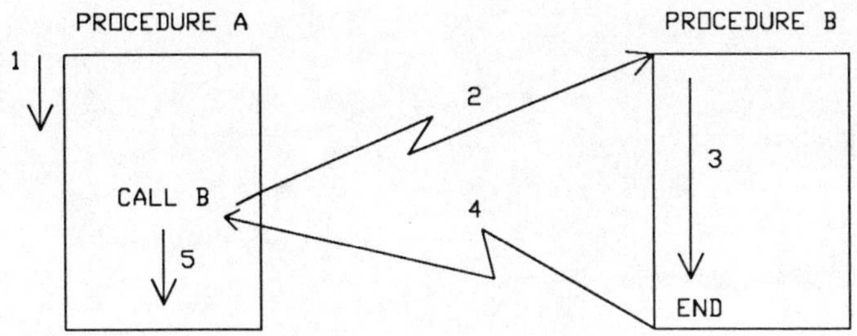

Figure 2-1. Procedure Call.

and proper use of passed information, we can construct an effective resource-sharing mechanism and therefore effective user applications.

Remote-procedure call (RPC) mechanisms operate like normal procedure calls; that is, the caller issues the call to a known procedure, the caller goes into a wait state, the remote procedure is begun, the remote procedure completes, and returns control to the caller's machine. The RPC mechanism wakes up the caller and allows it to continue (Figure 2-1).

The RPC mechanism uses a master/slave type of relationship to control how the computation ensues. In the above example, Procedure B has no choice. It is the slave of procedure A, so when it's called, it has no choice but to do as it is told. As in traditional procedure calls, the RPC can have parameters passed in as part of its structure. This scheme will work only if we have a reliable communications mechanism supporting it. This method of procedure interaction limits one as to the ways in which one can build distributed programs.

Many systems with RPC mechanisms have been built, since it is one of the simplest forms of remote control to implement and utilize.

Message passing is another popular way to build distributed programs. This scheme typically uses mailboxes or ports to which messages can be sent. Message-passing schemes do not require that the calling procedure or its intended receiver be halted or interrupted to accept the message. Message systems allow both to send and receive at will—which is much different from the RPC, where the master and slave must be tightly linked to perform their tasks. Message-passing schemes typically operate as follows:

The sending process (A) wishes to send a message to procedure B. It forms the message, addresses it to B, and places it in its box to be picked up. The communications facility sees that A's flag is up (something to send), so it extracts the message and sends it to the addressed process B. Process B gets the message in its mailbox. When procedure B gets around to it (Read B), it will examine its mailbox contents and decide what it will do (Figure 2-2).

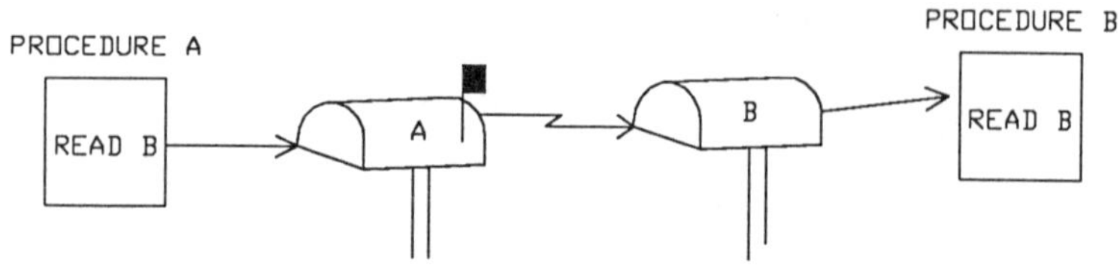

Figure 2-2. Message Passing System.

This is not a master/slave condition as in the RPC; instead, message passing is a cooperative venture where each process decides, based on predetermined protocols how to handle and interpret messages sent to it.

The third form of remote interaction utilizes the object model paradigm. In this case, procedures within the system are encased into constructs called objects. Objects are the granularity of item recognizable by the system and are the only means of interaction. Objects are viewed as abstract data types that have distinct operations (methods) that can be performed on them. Additionally, by definition an object cannot be acted on by any means other than its allowable operations. No other operations or actions have meaning to the object. Objects are protected by access tickets called capabilities. User objects (procedures) can only access an object if they possess a capability for it.

Capabilities provide objects with addresses and invocation rights on named objects that they are cognizant of. Object invocation and interaction proceeds as follows:

If object A possesses a capability for object B, and it wishes to invoke operation b on object B, it does this by providing the capability for B and the operation b in an invocation packet. These are handled by the low-level IPC to invoke the needed operation on object B (Figure 2-3). Object B receives the invocation operations request and initiates the wanted operation. Once the operation is completed, object B returns status of the operation along with any side effect (such as data structures, etc.) to the invoking object A.

Resource Server

Using these basic interprocess invocation facilities we can construct the next level of user applications support software, the resource server. As was the case with the basic mechanisms of invocation, so are there varieties of resource servers. Resource servers can be simple client/server types, or can be more complex distributed servers. Additionally, each of these can be constructed using one of the previously defined invocation schemes.

The client/server model (Figure 2-4) operates as a cooperative interaction between the client and the server. Clients request service from the server. The server acts like a traffic cop deciding who goes through to get service and how. Interaction with the

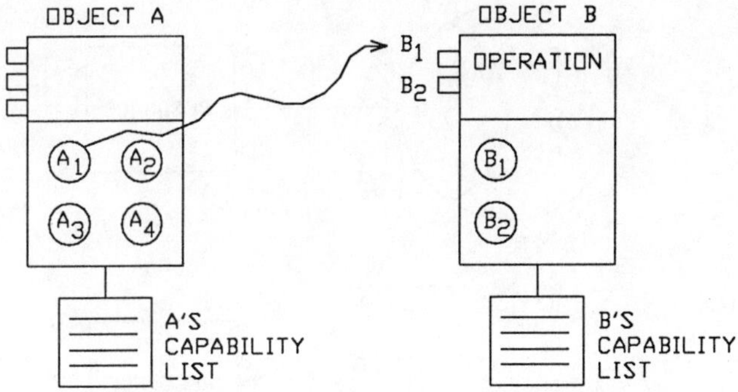

Figure 2-3. Object Oriented Invocation.

resource is handled by the server isolating or hiding the particulars of this action from the client. This form of information hiding provides the user with a clean and clear view of resources without needing to know about the network.

Servers are associated with each resource in the system, whether it be hardware or software, and provide for its ultimate controlled use.

The distributed server is a similar construct, except that instead of a server per resource, there is one server for each active client in the system (Figure 2-5). In this model, a client has a systemwide program copied into its program space that provides access to the resources in the system. This program provides the client with its interface to the server pocess. In either case, the details of actual interaction with resources are hidden from view and controlled by the server.

Servers are typically classified as monitors, managers, or distributed managers. A monitor server provides service for a resource that can be accessed only in a serial manner. For example, a tape drive or printer server would be constructed as a monitor server. A manager server would be provided to resources that can be acted on in a concurrent fashion such as disk drive banks, memory banks, special purpose processors, or other such concurrently accessible devices.

The distributed server is composed of a set of servers spread out over the LAN that provides a class of service. For example, a distributed database server is composed of

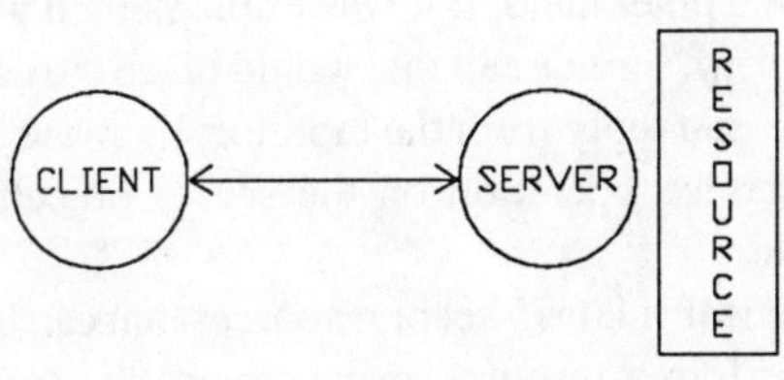

Figure 2-4. Client/Server Model.

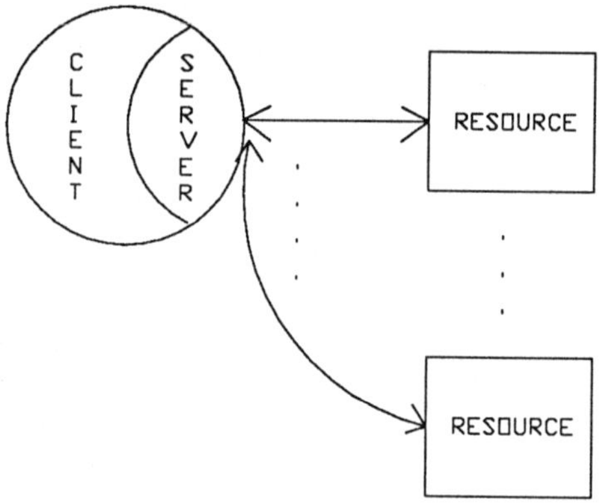

Figure 2-5. Distributed Server Model.

a collection of servers that together provide a unified concurrent and consistent database to users.

Monitor processes are used for resources that can be accessed by only one process at a time. The basic concept of a monitor server comes from Hoare [Hoare 1974]. The main point of the monitor server is that the sequence of actions on the controlled resource must be strictly serial. The skeleton of a monitor server is shown below:

```
Process monitor (x)
 Initialize
  While (condition true)
   Select next
   DO service
   Reply
  End while
 End monitor
```

From a user's perspective (a client), if this were implemented in the message-passing process-based model structure, then the client would need to issue a request service message to the monitor with appropriate conditions. If the monitor is free, it services the user. If not, it will queue up the user requests (done in its mailbox) and select them in order of arrival. On the other hand, if it was implemented with the RPC model, the client would issue a perform service call that would place the user request into a queue, then wait on the service-end reply from the monitor; likewise in the object model, the client would issue a service operation on the server object, then wait for a reply operation to continue on.

The manager server encapsulates a set of resources that can be offered concurrently. The manager server performs resource management by cooperating with a set of associated workers (Figure 2-6).

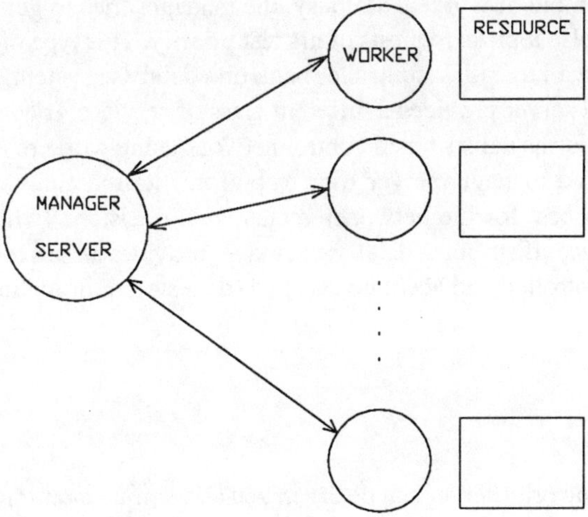

Figure 2-6. Manager Server.

All resource manipulations are handled by the workers with coordination of access provided by the manager. The relationship between the manager and workers is that of master and slave, where the slaves do one job and the master coordinates the actions of many. A simple skeleton of a manager is shown below:

```
Manager
  initialize
  while (true) do
    tag= all workers busy? worktag:max
    PID= get worker not busy
    switch (acquire message from worker (PID)
  Case worker:
    update worker status (PID and message)
    if(waiting reply)
    reply to waiting process
    break
  case user
    enqueue (request,PID,message)
    break
 if (not qempty and not busy worker)
  reply worker (first (request))
 End
```

The manager can receive a message either from a user process or its associated workers. The manager controls the incoming request messages based on their message qualities. For example, if workers are busy, the manager tries to get a message from any worker first, else it picks requests of highest priority. This type of manager server would be found in a large disk-controller bank or a database system.

The distributed server provides a different class of services. It consists of a group of managers that forms a distributed control network using workers from the managers. The distributed manager server tries to perform its functions based on global decisions (what's best for the network) versus local decisions (what's best for my node). For example, a distributed database manager provides a class of service to users. It provides for controlled and accurate access to data stored throughout the network.

Examples

Print Server The print server can be organized as a simple monitor or as a manager server. It controls a printer or printer's resource and offers three simple services to any number of requesters: open, print, and close. To print a message, file, character, etc., on the paper, a client first must send an open request (to acquire buffers for storage), wait for a response on request, then send a print request for each line to be printed, and upon completion issue a close request. This type of operation is a simple line-at-a-time print server; we could instead allocate more buffers and provide a server that does an entire file at a time before it responds. As long as all the clients obey this "protocol," the monitor server guarantees that printed outputs from one client will not be interleaved with outputs from any other clients.

If the printer is in use when an open request arrives, the print server defers the open request until the printer is free and all previously deferred open requests have been met. That is, the arriving requests are queued up and served in FIFO order (first in first out). Conversely, if the printer is free when an open request arrives, the print server responds to the client's request, enabling the client to start issuing print requests. Print requests are acted on and set only by a client who has been issued an open request acknowledgement. Once finished with its printing, the client issues a close request, which results in freeing of the printer and the removal of a pending open request to print the next file.

This process is shown below:

```
Process print server
 begin
  printing = false
  do
   receive (message);
   case message of service:
     open:
      if printing then
       enqueue (message, open Q)
```

```
    else begin
     printing:=true
     respond (message)
    end
  print:
   begin
      If open(client) then
   print (message);
   respond (message)
      Else(error)
   end
  close:
     If Open(client) then
  printing;=false
     Else(error)
  if queue not empty dequeue (open Q);
  response (message)
  end
 until false
end
```

This process could be made more client-friendly by providing variable print sizes versus single-message-at-a-time printing. We could have it accept files of variable size, from a single character to hundreds of pages, for example.

We could also alter this process to become a management server by providing it with N monitor servers and code to manage them. The added code would be quite simple—just an additional data structure to maintain a list of free print servers and code to ship work to whichever one is free:

```
process manager print server
begin
 set all not busy
 do
  receive (message)
 case message of request
 open;
  for i = first to last printer do
   if not in use then message.printer=i exit loop
   if in use then in use [i]=true;
     message.printer:=i
     respond (message)
  else enque (message,open q)
 end
 Print:
```

```
  forward to printserver (message, ID, message.printer)
 close:
  inuse[message.printer]=false
  deque (openq)
  respond (message)
until false
end
```

Additionally, if we wished to have printers spread throughout the network and control them by some intelligent means, we could construct a distributed print server. This server would be a collection of managers, possibly for a variety of printing devices such as line printers, letter-quality printers, laser printers, large graphics printers, color plotters, and many more. The managers could be organized as a collection of managers each for a special class of printing device. These managers would manage their pool of printers/plotters, etc., for the distributed server.

Clients would send requests to the distributed server on their node in a logical English form, for example:

```
Plot(space shuttle.dwg,print-quality)
```

which would indicate to the distributed manager that a graphics plot of the graph file spaceshuttle.dwg (a drawing of the space shuttle) is to be done. The distributed print server would decide, based on available equipment and quality of plot requested, which device to print it on. The user need not know or care where it is being done so long as he gets it back in the requested form through company couriers or any other means. If a user wishes to receive the plot back immediately, then he or she must request a plotter known to be physically close in proximity.

Tape Server

A tape server also can be constructed in any of the three ways described above for the print server. It, too, is a sequential access device and can perform only a few simple tasks. In the basic form it can be read, written, or rewound—nothing else. Additionally for management purposes, we must have synchronization controls; in particular, we will use "acquire" and "release" to indicate getting in for service and releasing from service. In the simplest case, where we force the user processes, objects, etc., to have knowledge of how to use the device (i.e., they must know that the protocol for use is acquire first and then perform any of the other three: read, write, rewind, then release upon completion).

```
Process tapeserver
 Use=false
 do
```

```
receive(message)
case message of service
acquire
 if use then
   enqueue (message, acquire Q)
 else
   use=true
   respond (message)
 end
 Read:
   read (message.fname)
   respond (message)
 end
 write
   scan until EOF found
   write (message.fname)
   respond (message)
end
rewind:
 rewind (message.location)
 respond (message)
end
release:
 use=false
 dequeued(acquireQ)
 respond (message)
until false
  end;
```

This simple procedure will allow any number of clients to line up for use of the tape resource, but only one at a time to actively read, write, or rewind the tape. As in the printer case, we can easily transform this into a manager tape server by allowing multiple clients concurrent use to the maximum number of available tape drives. It gets a little sticky, though, when we must know which tape a file is on. In this case, the tape server would need to have a dictionary service provided that lists the contents of the available loaded tapes on the various drives.

```
process manager tape server
 with available tape drives do
  inuse=false
 Do
  Receive(message)
  Case message of service
 acquire:
```

```
 if use (funit[message.fname])then
  enqueue(message,acquireQ)
else
  acquire (fname.Tunit)
  use[t-unit]=true
  norespond(message)
end
read:
 read(t-unit,message.fname)
 respond(message)
end

write:
 scan until EOF found
 write(message.fname)
 respond(message)
end

Rewind:
 rewind(t-unit,message.location)
 respond(message)
end
Release:
 use[t-unit]=false
 dequeue(acquireQ)
 respond(message)
end
until false
 end
```

This pseudo code describes a manager server for the tape units. It will provide a way to access a particular drive by filename (stored in a directory) and allow write access to any available unit. User would access the manager by providing a command to acquire (message.true) and then perform the intended function of read, write, rewind.

The distributed tape server would have the same structure as the distributed print server of the previous section. It would be used to provide transparent service to clients throughout the network. Clients wishing to use a variety of tape storage and retrieval devices would issue simple commands to read/write a file, and the distributed tape manager would provide the transparent service to acquire a tape drive of sufficient power and, possibly with the proper file, to control the interaction for the client.

Disk File Server

The disk server can also be easily constructed using the same techniques. That is, it could be built as a monitor (for each disk drive), as a manager (with N monitors, one for each drive), or as a distributed disk server. In each case, the structure of the server would be similar to that of the previous servers except that it would be a more concurrently available device and therefore would need some additional services to handle this condition. Clients would not need to acquire the drive. Instead, they would need to open a file to read it or write to it, and close it once they are done. This feature would allow the disk server to provide service to as many clients as need it, as long as they are not accessing the same file. Clients would access the disk server by specifying an open (filename) command, then issuing the appropriate read/write, append to this file, and once completed issue a close (filename) to free the named file for other clients to use. This feature guarantees the nonconflicting use of the files while providing better overall concurrent use of the device. A simple example of a monitor disk file server is shown below:

```
Process monitor-disk-server
Receive(message)
case message of service
open:
  if file open already
   then queue (message)
  else
   set file open flag
   respond (message)
end
read:
  read (filename)
  respond(message)
end
write:

write(filename)
  respond(message)
end
append:
  append(filename)
  respond(message)
end
close:
  set file open flag to closed
```

```
  if queue(filename)notempty then
   deque(openq)
  end
end
```

This simple example provides a means for many clients to access this disk device and acquire files for use. One could include additional logic to check clients' authorization to make sure they are allowed to access a file for read, write, append, all or some combination of these rights. As in the first example, this server could be implemented using either the RPC, object, or in this case, process message-passing model of interaction. Like the print server and tape server, the disk server can be constructed as a manager server. In reality, most disk systems are organized this way, even on many of the newer personal computers and workstations.

A disk server built as a manager would provide the means to control the use of a bank of disk drives, which is the normal mode in most minicomputer, mainframe, or supercomputer systems. The manager would accept requests to open files, read, write, append, and close files by name only—not by a device number or identifier. The manager would have the job of determining where the file was stored and providing the client's requested service. As in the simple example, reads, writes, and appends are allowed to clients who have opened a file. A file can be opened by only one client at a time.

The distributed disk server would provide clients with access to disks residing on a variety of machines and with varying speed and performance characteristics. It would provide transparent access to the devices on the physical level and perform all updates, etc., to keep the possible multiple copies of the disk data files updated correctly.

Combined Distributed Server

A more interesting server would be a combined server that provides three classes of service to users: storing and retrieving files from a random access device (disks), storage and retrieval from archival devices (tape servers), and printing of files (print server). This distributed server would exist on all nodes and would provide distributed access and control to the variety of devices. Now we're getting to real resource sharing in a distributed system. The distributed server would have asked as follows:

```
Distributed File Server
 Receive (message)        get request from users
 with message.service do
  if print then
   print (message)        send to print manager
 else if tape then
  tape (message)          send request to tape manager
```

```
else if disk then
  disk(message)          send request to disk manager
end
```

This is a very simplistic notion of how this function is actually accomplished, but I think you get the idea. The distributed server would provide a single place to direct requests to, leaving all details of the actual interactions with the low-level service managers.

DISTRIBUTED PROCESSING

Users can construct their own resource-sharing systems using the basic invocation mechanisms. They would use either message passing, remote procedure calls, or object invocations to construct various protocols to provide synchronization of distributed segments of their intended distributed application. For example, in a message-passing scheme, mailboxes can be used as a synchronization variable (semaphore). Clients wishing to cooperate with each other in doing some tasks—for example, the producer/consumer problem—would use the placing and removal of items from their box as their signal and wait semaphore conditions. There are two ways for the clients to view the mailbox; in one, synchronous, the client places the message at the box and waits for the other to come and remove the message. In asynchronous, the client places the message in the box, then continues its processing. Using the second case, we can build a remote producer/consumer relationship as follows:

Client A and Client B each share a mailbox M. Client A can only put items into the mailbox. Client B can only extract them. Client A is the producer, and produces messages P continually and places them in M until M is full, at which point Client A performs a busy wait (continually checks status of mailbox M until such time as it has had items removed). Client B performs a busy wait on an empty mailbox. When it sees items in M, it removes them one at a time until the box is empty, at which point it goes back to a busy wait (Figure 2-7). The mailbox facility itself performs the mutual exclusion by accepting and allowing removal in order of occurrence. Mutual exclusion is the condition of allowing only one user task into the shared message box at a time.

We could also construct this using the object model, as follows:

Object A is the producer, object B is the consumer, object C is a semaphore that allows wait and signal on two held variables (S,S1).

```
producer object a
 wait(s) object C
   produce item
    count-count+1
  signal Sl object C
```

```
consumer object B
 wait Sl object C
  consume item from buffer
  count=count-1
  release 1
 signal S object C
```

The producer waits for object B to signal that an item has been taken and then adds an item, signaling to object B that an item is there to be taken. Object B checks to see if a signal has been issued; if so, it removes an item and then signals to object A that it has removed an item.

Using this notion of wait and signal in either start sequence or asynchronously allows the clients to construct distributed applications. Details of other features of distributed processing and applications synchronization can be found in Fortier [1986] and Desrochers [1987].

MODEL/SIMULATE/FORECAST/PLANNING SOFTWARE

The previous section introduced some of the basic applications facilities and services. These facilities and services are needed so that additional networking applications can be developed and implemented.

This section examines the capabilities and concepts associated with LAN software aimed at assisting corporate planners as well as researchers. In particular, we examine some of the basic capabilities we would like to see in LAN software within this category.

Modeling is an important feature of most large business applications for management as well as for research. Models are used to construct representations of physical systems to allow study beyond what is possible on the real systems. Models allow for studying systems in expanded time, compressed time, and real time. They allow the stress testing of critical systems as well as sensitivity testing on system components. Models can take on different forms. They can be analytical, simulation, or operational

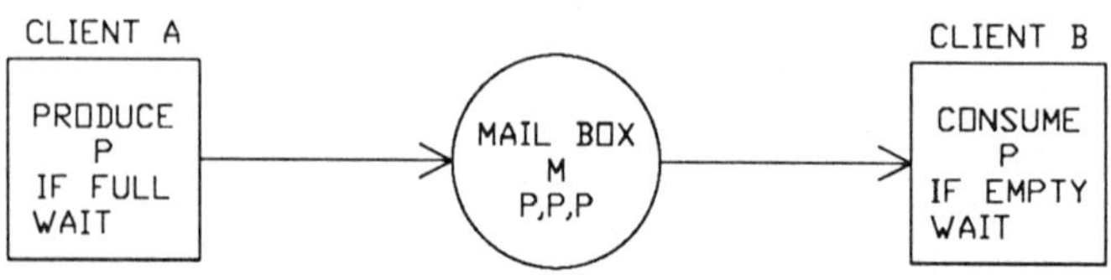

Figure 2-7. Producer/Consumer Message Passing Model.

models. Each has something different to offer to the corporate user. The analytical models provide a means to build representations of static systems rather quickly and give good results in terms of average and worst-case conditions. Typical of analytical models are queuing analysis models. Simulation models represent the next class of modeling tools. Simulation provides a tool that can model at varying levels of detail and permits experimentation with potentially volatile systems without touching them. It forms an artificial laboratory. Simulations come in a variety of flavors, covering a discrete event, continuous, combined, queuing, and hybrids.

Discrete event models require the modeler to construct each component of a system under study as an event, and to define on event boundaries how the state of the various components can change. Continuous modeling relies on the description of the system components and interactions using continuous formulas. These formulas describe how the variables defining the system change in relation to time. A definition of crossing conditions and other factors can be used to adjust the formulas. In this fashion a wide range of modeling is possible.

Queuing models provide a simple means to construct queuing models of a system and study the effect of various conditions on the system. Hybrid models give a way to combine the features of each and use the best aspects of each in modeling a system.

Simulation models of many systems have been built and can be purchased by the LAN user. They provide generic services to network users to access them and to construct models for use in

- Forecasting
- Weather
- Economic conditions
- Financial conditions
- Manufacturing

These results can then be combined with others to provide tools for planning an enterprise's actions. Network-based tools for these areas are few and typically found only on large machines today, although the trend is changing.

The last category of LAN software within the present heading is data processing software. Typical of this class of software are applications requiring data reduction, interpolation, and extrapolation. This software provides services to the other simulations, forecasters, and planning packages. Sensors can be strewn about the system or within a plant, for example, and data collected, reduced, and processed to provide a wide array of data for bussiness planners to utilize for analysis.

SOFTWARE ENGINEERING ENVIRONMENTS

The role of this category of software is to provide design, development, maintenance, and execution support for all applications software within the local area network. Most of us will recognize these as part of the latest industry buzz word, software-engineering-environment (SEE). In the context of a SEE, we are interested in requirements analysis tools, specification tools, design aids, implementation aids, testing aids, and maintenance and documentation aids.

The key to calling it a SEE versus uncoupled tools is the integration via a unified database and schema. The database is the integrating component and will be discussed in general terms in the next chapter. The development of software engineering environments has been pursued for many reasons:

Shortcomings of Current Software Development Practices

- Inconsistent, incoherent, incomplete designs.
- Responsibilities, duties, and accountability are poorly defined and controlled.
- No provision to perform impact analysis in relation to specific parameters.
- Completion criteria are poorly defined.
- Histories are unavailable.
- No metrics are available to test goodness.
- No effective means to manage complexity.
- Unreusable software and software designs.
- Designs fail to provide for change.
- Ineffective communication/feedback.
- "Team" efforts are not coordinated.
- Performance assessment done after the fact.
- No effective documentation aids.
- No assistance available for the resolution of conflicts or issues.

Resulting in:

- Excessive development costs.
- Unreliable and unmaintainable products.

Based on these shortcomings, many agencies such as the U.S. Government (STARS), foreign governments (ALVI, Espirit), and many companies and universities (Software Engineering Institute, SEI, at Carnegie-Mellon University) have initiated research and development efforts to develop integrated environments that:

- Support the entire software life cycle.
- Are methodology driven.
- Provide capabilities flexible enough for all phases of the software lifecycle. (See Figure 2-8.)

The software engineering environment is an aggregate of surrounding things, conditions, and influences that affect the development of a software program. A software engineering environment consists of the utility programs, operating system, file systems, languages, compilers, assemblers, interpreters, editors, debuggers, and

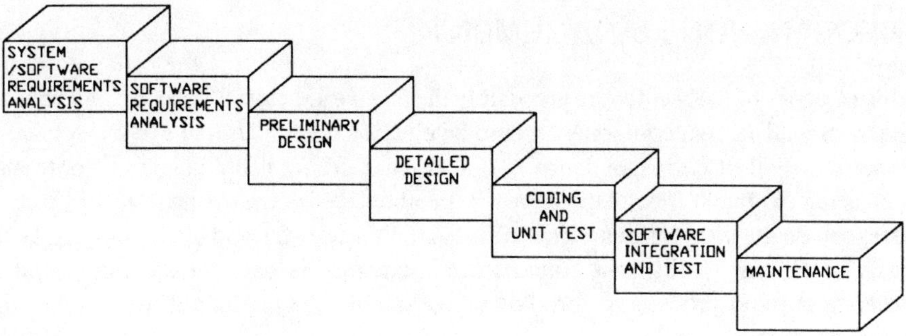

Figure 2-8. Software Lifecycle.

development tools. All of these tools are used to enhance and improve the quality and quantity of software produced. The heart of a software engineering environment is its methodology. The methodology is the process by which software is developed. A methodology must cover the entire life cycle of software development (Figure 2-8). It must be able to successfully produce and transform system requirements into a precise statement of the software system's external behavior. It should possess a creative aspect, one that can help derive the specifications, as well as an easily used clerical aspect to document them. Moreover, the method must fully describe the interfaces, modes of operation, and functions of the system. Important characteristics [Charette 1986] are that it produce blackbox views, be understandable, accurate, precise, and easily adaptable; it must support the realization of the system being specified in some abstract form. The methodology must support abstraction, decomposition, notion of hierarchy, and provide for traceability and correctness analysis. The creative and clerical aspects of the method must be supported by techniques such as information hiding, abstraction of data types, stepwise refinement, data flow analysis, and graphic decomposition.

These represent the features one would like to have in a complete methodology, although in reality we typically see some subset of these features being realized. Some examples of methodologies include Yourdon, Inc., Structured Systems Analysis [Demarco 1979]; SRI Hierarchical Design Methodology [Robinson 1979]; Soem [Charette 1986]; The Software Development System [Davis 1977]; User Software Engineering [Wassermann 1979]; etc.

From a LAN perspective we wish to have an environment that lends itself to the development of distributed software programs that can operate in our system. This requires tools tuned to capturing the design of the software but also to the topology and structures of the system it will run on. Presently there are not many languages designed specifically for LANs. Even beyond this there are not many compilers or tools that have been specifically designed to aid the LAN software developer, though they will appear as time goes on and user demands fuel the fire. Present languages and environments are mostly research oriented.

INFORMATION MANAGEMENT

This category of LAN software is possibly the one we are most familiar with. Probably the most well-known component would be electronic mail. Users have the capability to send E-mail or correspondence to one another. Many people use electronic mail versus regular mail for correspondence with other users of the media. The information category deals with various forms of information transfer and presentation. At the basis of all the applications found in this category is data transfer, presentation preparation, and processing for showing. Examples of products found within this category include:

- Graphics
- Memoranda/letters
- Bulletin boards
- Report, specification, planning
- Teleconferencing
- Library searching
- Database servers
- Facsimile (FAX)

Graphics

LAN software involved with graphics comes in two major flavors of graphics designing and network transmission. In the design area, users who wish to create graphics for shipment over the LAN or for internal use can either use canned cut-and-paste type programs, which allow users to select icons (pieces of a picture), scale them to desired sizes, and place them on the screen where they are to reside in the final drawing. The second style is freehand, where the computer allows the user either via keyboard commands or an attached device such as a light pen, mouse, digitizer pad, etc., to draw lines as if holding a pencil. This method allows great versatility in drawing, but the user is left to his or her own artistic abilities. Beyond these two basic styles of graphic development there is a third, and it uses the best of both of these techniques. The graphics designer can select canned shapes and combine or change them via freehand actions. The result is graphics that have the freehand quality in much less time than the purely freehand.

To communicate drawings, the LAN provides two mechanisms to send the data. The data can be sent as straight bit maps or as coded versions of the same. In the uncoded format, each bit on the screen is sent over to the other device that is to show this image. The volume of data can be quite large for a small screen, in the vicinity of 500K bits; with coded formats, the only bits on the screen providing information are coded (address/intensity) and sent over the media. This coded method results in less data being sent (reduction of 50% or more in some cases), but at a cost of coding and decoding on the sender and receiver side. Other languages, like postcript, encapsulate drawings as machine and resolution independent procedures that can then be transmitted to any device possessing the ability to translate them.

The second category of information applications is the memo or letters category. The applications in this category deal with the production, shipping, and delivery of memos and other small correspondences among network users. Typical of applications of this type are mailer packages that provide users a means to send files or just-created script to users, using a hierarchical addressing scheme such as

username@computer.localareanetwork.wideareanetwork

The mail or letter is then sent over the networks named to the end user addressed. At his site it is placed in a mail file for the user to view when he gets around to it.

A third category, the bulletin board, is similar to the mail packages except it sends its messages to a globally available file where all users of the network can view them. This differs from the mail system that places the letter in a place only viewable by the end user to whom the letter is addressed, and by no one else.

A fourth category is report generators, which provide a means to generate reports, text and graphics, ship them to others for review, and receive them back again. Required by applications of this type would be transmission capability for large blocks of data, and creation and editing capabilities for the file. Additionally, this type of application would require the maintenance of versions of documents for recovery to old versions, or to incorporate only some comments from reviewers and not others. It would require as part of its editor suite, cut-and-paste capabilities to allow editors to extract pieces of one document and place them in another.

A newer and quickly expanding use of LANs is application teleconferencing. Many companies are installing video-quality LANs (high bandwidth/speed) to allow the construction of facilities by which they can have conferences without the travel and time associated with it. Parties who wish to meet schedule time at their prospective centers, possibly via E-mail, and then get together electronically at that time. Teleconferencing allows the individuals at both sites to see each other, hear each other, and receive a number of other screens with any information required of the meeting (such as viewgraphs, charts, papers, etc.).

LANs to support this class of information applications must be high speed and have high data capacity.

Library Searching

LANs have also found their way into the library. Many large libraries now offer online access to a variety of databases. One can simply log on to the library of interest, formulate search queries, and let the computer do the work for you. Typical of such systems would be a set of workstations connected over a local area network with possibly multiple servers with access to outside wide area networks and remote databases.

Database Servers

Finally, under the information category are database servers. Many various services have been developed where one simply logs into a remote database and then can access a variety of information. As more databases arise, more LAN vendors will offer software that can access, process, and inform the LAN users.

DESIGN

This category of LAN software applications deals with the problem of design. Computers have been increasingly used in the design process for all aspects of industry; e.g., publications, drafting, machine design, architecture, VLSI devices, automobiles, aircraft, spacecraft, and many more. The computer is becoming an increasingly important piece of the process and as time progresses, so is the LAN. LANs provide the means for multiple CAD workstations to share the pieces of a total design while keeping it as a single whole. The LAN provides a means for designers to examine each other's efforts and use them in future designs. CAD tools such as AutoCAD from Autodesk provide PC-based design aids for a variety of uses. The LAN can provide a means for the CAD system to access large data banks of past designs that could be used to simplify the present job. The LAN additionally provides the CAD user with access to potentially higher-powered design tools, such as plotters, graphic engines, simulators and many others.

CAM (computer aided manufacturing) utilizes designs generated by CAD systems to operate various milling and manufacturing machines in the fabrication of products. The LAN is used to link the various robotic devices together, allowing the LAN system to set up and control the production process; data flows from device to device, providing for synchronization of activities and notification of conditions. An example CAM facility and LAN network is described in Komoda, 1984. This system is composed of multiple ring topologies interconnecting various sensors, robotic equipments, workstations, and peripherals configured to provide an autonomous decentralized control system for factory automation.

CONTROL

An example of a computer-aided information management and control application is teaching. LANs and computers have been finding their way into the classroom. Packages for teaching students with learning disabilities, for gifted children, and just to teach computer literacy have been growing in demand and presence in the classroom. LANs provide a means for schools to link teaching devices to a teacher's control computer that can provide the type of learning tools and situations needed by a child. This type of application can provide future gains in the quality of instruction in our schools.

This category of LAN software applications deals with control of remote devices. LAN software for simulators and team training for crews of high-technology aircraft, ships, and spacecraft have been developed. In the future, similar software for other systems will ensue.

Control software can be used to automatically control processes, as would be the case in a closed-loop control environment such as a power plant, or for interactive control in situations where human intervention is needed for safety or performance. Generic in this type of LAN software are components to sample data throughout the system on periodic and aperiodic bases, to condense the sampled data into composite views of systems activity, and software to analize and assess the impact of failures and to effect corrections based on the sensed data.

SUMMARY

This chapter reviewed the categories of LAN Applications Software in generic terms, leaving details of actual products for later chapters. The goal of the chapter was to introduce the reader to the way in which a LAN is used by the various applications and how a LAN improves the application's and user's performance.

3

SYSTEMS MANAGEMENT

INTRODUCTION

Computer systems, whether distributed, federated, or centralized, require some form of control. By control we mean orderly interaction of users in their utilization of systems assets. This implies fair access to these assets and guarantees against lockouts and failures due to user actions. The control software must be able to accept requests for service, determine what to do, and act accordingly. It must be able to force interrupts (to intervene on an erroneous process), allocate CPU time evenly, and act on interrupts from other processes. In short, the control environment is the traffic cop. It directs all the actions of the involved entities providing for smooth service. A LAN control environment is no different except that it performs its function over many devices located in possibly distant places versus a centralized computer's tightly coupled peripherals and local processes. A LAN control environment must provide services that guarantee security (network access restrictions), provide transparent communications services, provide global information management, and provide for the controlled interaction of users' code.

LAN OPERATING SYSTEMS

Operating systems for local area networks come in two basic flavors, network operating system (NOS), and the distributed (global) operating system.

Network operating systems provide services and mechanisms for local operating systems that provide a network view. A network operating system is one built on top of and around a local operating system (Figure 3-1). The user processes within this

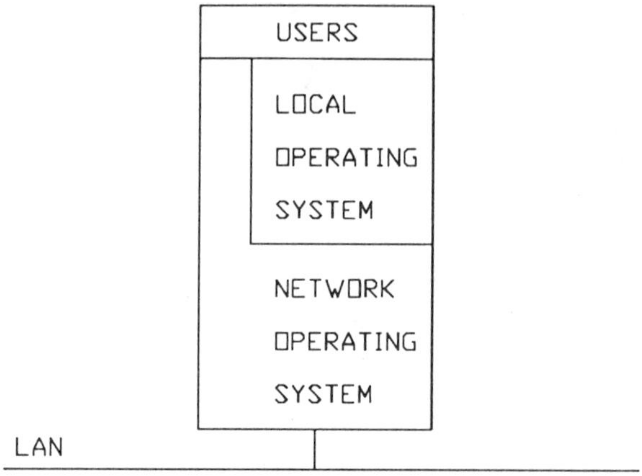

Figure 3-1. Local Operating System Structure.

type of architecture must be aware that a network exists, and possibly even know how to access the remote servers needed. The NOS is constructed to serve the local operating system by providing interfaces between the remote local operating system and the local one. It additionally provides mechanisms for user processes to converse and synchronize with each other over the network.

The second form of operating system for a local area network is the global operating system. This form of operating system is one built from scratch with the network in mind. The basic differentiating feature of a global operating system versus network operating system is that there are no local operating systems (Figure 3-2), and users have a seamless view of the provided services.

The distributed operating system is built to handle both local and global resource management from the perspective that the collection of nodes and resources represents one system to manage versus N distant ones. A distributed operating system is built on top of the bare machine. It is not an add-on to an existing system's software. The distributed operating system manages all resources without contention from local service software.

The global operating system covers:

- Process management
- Memory management
- I/O management
- Device management
- Network management
- File management

Within this type of architecture each component manages resources at a global level. That is, they provide services that act to ensure "total" systems operabil-

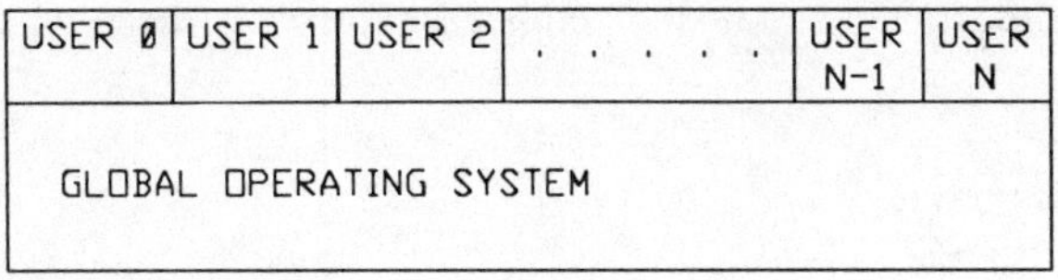

Figure 3-2. Global Operating Systems Structure.

ity/performance, possibly even at a cost of local performance. Decisions on management are based on global data and are performed by cooperation and consensus.

Whether a LAN operating system is of the communications server (NOS), network, or distributed type, they all have been constructed for the same purpose—that of providing general solutions to user problems associated with managing their applications. The operating system isolates the user from dealing directly with the intricacies of the system resources that he needs to utilize.

The operating system's communications software removes the details of message formatting, addressing, network access, transmission, and error detection/correction from the user. The operating system provides mechanisms that implement the policies established by the designers for the sharing of systems resources among users. The process for sharing and coordinating can take on various forms. For example, we could utilize a set of servers for each resource and require that they coordinate with each other in making decisions. The level of coordination will determine the extent of control as well as the performance. The coordination level directly affects performance. Since devices must exchange information to coordinate, we suffer some form of degradation of performance. Alternatively, we could construct a system where the resource managers cooperate, but in a much looser sense. That is, they make their decisions based not on active interaction with their peers, but instead via some best effort analysis, where they make decisions based on the best available data they have been compiling. In either case, decisions are made to provide transparent service to the user applications.

An operating system, whether centralized or distributed, must perform some basic tasks in order to manage resources, namely:

- Recognition that something wishes service.
- Mechanism to determine what to serve and where.
- Mechanism to provide service and take it away.
- Mechanism to detect exception conditions.
- Mechanism to allow synchronization.

Specific mechanisms implement concrete policies for determining how to provide service to a set of users. Given any set of policy conditions, a variety of mechanisms can be constructed to implement them. Policies in operating systems are composed of three parts:

1. A decision rule
2. A priority rule
3. An arbitration rule

That is, whether scheduling some service for a user, or determining what user to act on next, or how to provide synchronization, there exists a set of rules to describe how to perform the control act. The decision rules define when an operating system's operation is to occur—for example, preemptive or nonpreemptive operation (allow a task to be interrupted or not interrupted until fully completed). Each of these decision rules govern the feel of the operating system; that is, it defines its responsiveness and operations. The priority rule defines how an operating system will view its offered jobs. For example, users of class X get high priority (more cycles), and those of Y and Z low priority (less cycles). Priority defines orderings between managed resource operations. For example, in a disk drive, priority defines how the disk operates on its given request. If the disk is FIFO with no priority, then the jobs are accessed in sequence; if it has deadlines associated with accesses, then the queue can be organized by closest deadline first, longest first, etc. The point is that the priority rule determines how the offered work is to be viewed. The arbitration rule resolves resource conflicts that occur in systems. An arbitration rule tells us how to react when multiple requests of the same priority and service are encountered. For example, if we have contention, we may choose what to do based on the location in the queue or the requesting process semantics (its intent).

In either case, these rules are not implementations, but merely abstractions as to how the operation system components are to react to varying situations. Algorithms need to be developed, and code from them then implements the rules of control.

As was indicated earlier, the operating system must have policies and mechanisms to enforce the controlled use of resources. It must be able to detect that a process wishes to acquire service; it must possess means to determine what the service is and how and when to provide it. It must be able to detect and act on error conditions and must provide means for cooperative interaction with users; and among users, the means to realize these controls in the basics of entity management. Entities can be processes, objects, programs, procedures, etc. These represent the level of interaction schedulable and recognizable to the operating system. Anything below this is simple instruction execution; everything above has no meaning except in the eyes of the developer who designs coordination into his or her process.

Three basic models of operating system structure exist along with their attached policies/mechanisms. They are:

1. The process model
2. The object model
3. The remote procedure call model

The services required of an operating system and the means to carry them out can be physically realized by many means, although the three aforementioned models are the best-known from a LAN standpoint.

Object-Oriented Paradigm

The object paradigm relies on the structure of an object and its functional interaction to construct systems management services. The object model utilizes the object as the basic computational entity. That is, the object is the only component recognizable and accessible by users and management components. The object is defined as an abstract data type consisting of two principal parts:

1. An externally available specification
2. An internal realization

In generic terms, it consists of a specification and a body. The object encapsulates all aspects of its state and allows no access to them. The internal body part is invisible and inaccessible by any outside element. The object maintains its state entirely within its structure. This state can only be altered through the use of operations performed on the object. In addition, these operations represent the only means by which this object can be accessed and, therefore, its state altered. For example, a queue object would have operations to test if queue is full or empty, and to enqueue items and dequeue items. Nothing else makes sense on the simple queue object. You could not, for example, given this set of operations, push an item onto the queue, or pop an item off of the queue. Such instructions would have no meaning according to the definition of the queue object and therefore would be ignored. Access to objects on most object-oriented systems is controlled via an access ticket referred to as a capability. Capabilities for an object are granted to other objects that would like to use the object and have a right to it (by design or by structure). The holder of a capability for a named object has the right to access the object. Additionally the capability can have embedded its access rights that indicate what operations on the named object the invoker is allowed to access. A typical capability would look as shown in Figure 3-3.

The object identifier is used to either point directly to an object or point indirectly, through a table or conversion mechanism, to the intended object. The access-rights field contains a set of rights that this user has over the named object (for example, in the queue, one user may be granted the right to enqueue data but nothing else; whereas another would be given the right to dequeue data, and nothing else; and possibly a third user given the right to access all operations). Objects within this model of operating system design interact via the invocation of operations on their objects. For example, we could construct a producer/consumer system by using three objects: a producer object, a consumer object, and a queue object (Figure 3-4).

OBJECT ID	ACCES RIGHTS

Figure 3-3. Capability Format.

The producer object would possess a capability for the queue object with access rights of enqueue, full, empty granted to it. The consumer object would possess a capability for the queue object with access rights of dequeue and empty. The queue object is a simple resource object and has no capabilities to access any other objects in the system. The operation of the system would be as follows:

- The producer object tests the state of the queue object; if it is full, it exits and reinitiates the access (we assume a busy wait).
- If it sees the queue empty or less than full, it adds an item.
- On the other side, the consumer object accesses the queue object and tests if it has something; if so, it takes an item. If not, it busy-waits, waiting for an item to arrive.

This is a simple example, though it points out some of the means by which an operating system can be constructed using this model. The key element in making such a system operate is the proper definition of the interaction between objects and the controlled use of the capabilities for enforcing isolation. Operating systems of this type are described in Fortier [1986].

Process-Based Paradigm

In operating systems built for local area networks around the process-based model, all components of the active system are represented by processes. Processes are the active computationally known elements to the management software. All users exist in the system as processes and know of any other process identifier or association with a repository. Processes consist of a process body (the main code for the process), private

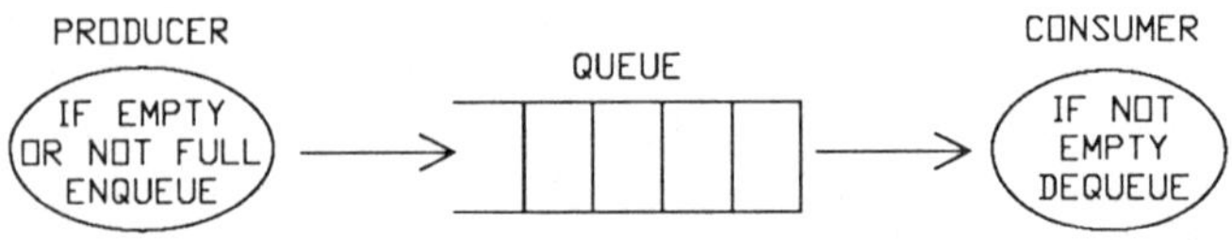

Figure 3-4. Producer/Consumer Objects.

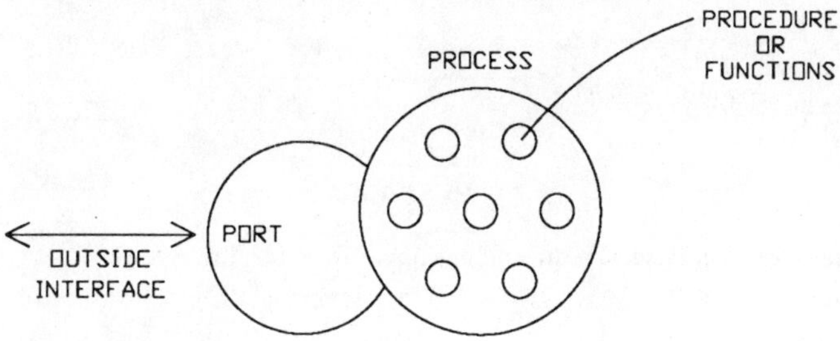

Figure 3-5. Process Structure.

data, and functions and procedures. The process encapsulates within its private address space all the entities comprising this process. These entities are not accessible directly from outside. They can be acted on only by the process interfaces provided. The structure of a process (Figure 3-5) provides a view of how this interface exists. Functions and procedures that are encased by the process converse with the outside world via a repository referred to as a port, mailbox, pipes, links, channels, etc., as they have been called. These structures provide a means by which outside processes can access these internal parts of the process.

The access, though, is dependent and controlled by the processes' internals, and at its own pace. There is no notion of interrupt or synchronization point. The outside sends messages to the port; the process, if it wishes to or when it wishes to, examines the message and decides what it will do. The basic notion here is that all communication and control occurs via the exchange of messages between process ports. No process is interacted with in any way except by message exchange. For example, using the producer/consumer model as we did in the object models case, we would have two processes, a producer and a consumer. Each of these would have a port associated with it, as seen in Figure 3-6. The producer process produces the product (whatever it is) and issues messages to port B on the consumer process to indicate it has produced something. The message is transmitted and is placed (stored) in the buffer areas of port B. The consumer process continually checks if any messages have been sent to it. When it sees a message in its buffers, it reads the message, sees that the producer has given it a product, consumes it (takes it out of its buffers), and sends a message to the producer to indicate it did consume the product. In this fashion we could build a producer/consumer procedure that toggles back and forth in a produce first, consume then reiterate fashion.

By the fashion in which processes examine their ports for messages and act on them, we can construct any form of resource management function that we wish. We could build a simple server, a multiple server, a manager, or any other type. The transportation of messages between the remote processes became the means to synchronize and

Figure 3-6. Process Based Producer/Consumer.

organize how the system manages the assets. Details of process-based operating systems structure, design, and examples can be found in Fortier [1986].

Remote Procedure Call

Operating systems built from this model have as their basic entities the program, procedure, and function. We all are, or should be, familiar with these basics of conventional programming languages. The interaction within this model is the procedure call. As in conventional systems, the operation is to call another procedure, possibly pass it some parameters, and then wait until a response is returned. This is a very simple mechanism and one can readily see how the client/server model can easily be fitted into this. The caller is the client and the callee is the server. The client initiates a request for service by forming a remote procedure call message containing the call and any necessary parameters. This message is sent by low-level interprocess communications mechanisms to the named process (the server), who handles it based on local protocol conventions. When completed, the server will cause a return from procedure to be performed, which will release the blocked caller and let processing continue. In principle, one can see that this model, with the proper support, can be used to construct an operating system environment for a local area network. The problem comes in how the operating system servers (spread out over the LAN) handle multiple, concurrent requests for service. The solutions are (1) let the user handle it, or (2) provide a systems service to handle the problem.

In the first case, the burden of how to handle this is put on each individual user, which causes multiple redundant and quite possibly conflicting solutions to be developed. The better solution is to provide operation systems administrator procedures that can queue up the multiple requests and in some orderly fashion (based on the policies in place) process the service requests. Additionally, the RPC-based operating system for a LAN must have a robust IPC mechanism to aid in localizing called procedures and for information transfer. The local components of the RPC-based LAN must possess the basic policies and mechanisms for procedure management (entrance, deletion, exit, resource access, interruptions, etc.), device management, I/O management, as well as many more. Details can be found in Chapter 7 of Fortier [1988].

DATABASE MANAGEMENT

A second major component of a LAN system management environment is the database manager. More and more systems are being delivered with database systems as a primary feature. The reasons deal mainly with industry's shift to information processing. That is, decisions in business, industry, government, and education are based on information. The best decisions occur when the best information is available. The way to provide the most up-to-date and accurate information to your user base is to install a local area network and a distributed database to manage the enterprise's information.

A database consists of many different components and, just as in the OSI/ISO reference model for networks, it is a layered architecture (Figure 3-7). The database manager consists of a user interface language, such as COBOL, SQL, QBF, that provides a means for the users to define data, and to access and process data.

This user interface provides basic database operations; that is, it provides a set of user queries (data access) and bundles them up into units of database consistency called transactions. These transactions are provided to the database query planner, which determines the content of the transaction and translates it into a system-usable form. This reduced set of access plans are then processed by an optimizer, which determines how to best perform the transaction's components to minimize the time it takes to perform the transaction and to maximize performance as measured by transactions per second. The output of the optimizer is an ordering of the basic database operations comprising the transactions. This execution plan is given to the Query

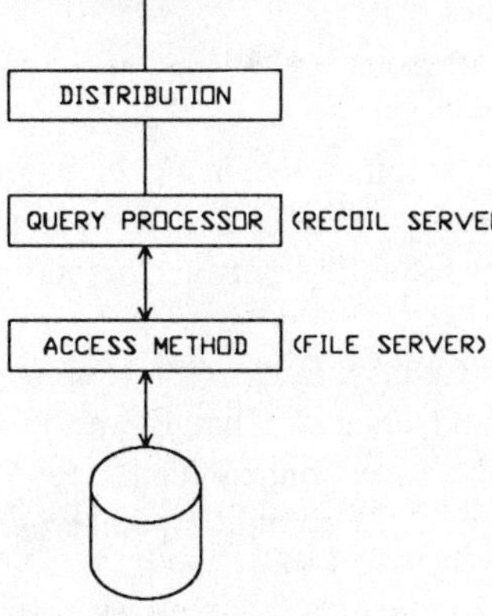

Figure 3-7. Database Manager Layered Architecture.

processor, which executes the Plan. During this execution, the query processor must perform processes to guarantee concurrency control, to keep data consistent, and to synchronize the access to other redundant or remote data.

Beyond this processing is simple file and buffer management software that controls the reads and writes of files stored on secondary storage. Each of these levels performs functions necessary to realize a working database manager.

The goal of a database management system is to model reality—that is, supply information necessary to model some set of real-world activities. What this implies is that the database must be able to model the relationships between data, not just the data itself. To provide this modeling of relationships, database systems have embraced one of four data modeling concepts, namely: Codasyl, relational, network, entity relationship.

These four models provide the means to construct data repositories and define relationships between this data. These basic data sets and the relationships form the model of reality the company wishes to capture. In all cases, the means of describing the basic elements and their relationships differ, as do the means of accessing this data. The differences lie mainly in terms of data structuring. For example, the relational data model provides a means to describe all data as tables (Figure 3-8). In the example of Figure 3-8, we wish to represent a university's course database. The pertinent data are the courses, instructors, and students. The relationships are who teaches what courses, where and when the course is offered, and what students are enrolled in this class. The relations given in Figure 3-8 are not optimized; that is, they are not normalized to the lowest level. We may have many duplications—redundant data that could be factored out if we used further reductions.

In this model to extract all students who took "CS 461" taught by Professor Ting, we would issue (in SQL) the following query:

```
Select students
    from STU
     where course#=
      select course #
        where coursename=cs461
```

or one of many other combinations. In any case, the "relationships" between data can be developed dynamically by the construction of queries. The penalty is that one must process more data items to extract pertinent relationships, but one has the added benefit of flexibility.

The hierarchical or Codasyl model requires more strict data structures, in which the relationships are "hand coded." That is, the relationships are built into the data structures. We lack flexibility, but can better describe and access relationships.

In the previous example, we could organize our data as a hierarchy of records, as shown in Figure 3-9.

The difference here is that we cannot ask questions not built into the data structure. At least not easily. We would have to write a program that scans each of the course

COU

COURSE NAME	COURSE #	CLASSROOM	SEMESTER

INST

INSTRUCTOR	COURSE #

STU

STUDENT	COURSE #	GPA

Figure 3-8. Relational Model of Courses.

name records to extract a list of instructors for courses, whereas in the relational case it already exists. This model is one of the oldest and is still used, although its popularity is waning as better-performing relational systems are developed.

Similar to the Codasyl data model is the network model. This, too, uses strict data linkages to describe the data structure and relationships. Similarly, it lacks flexibility to ask general queries but is efficient to hard-coded accesses. To access information, one navigates through the data structures accessing records until the one you want is found.

A more recent data model is the entity relationship model. This model uses features of the relational and network models in that it provides for general data tables as the storage entities and relationship links as means to indicate relationships. For example,

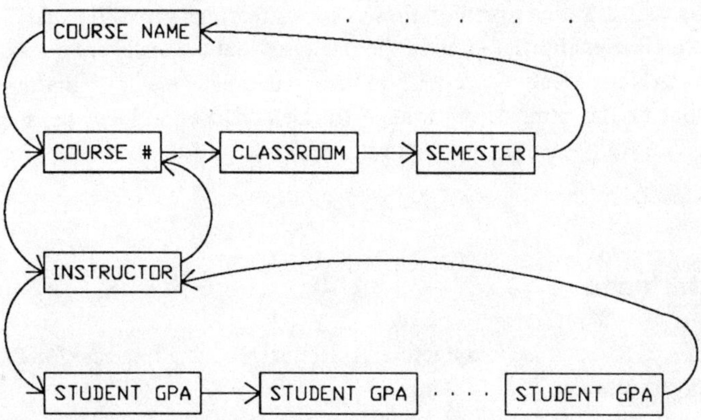

Figure 3-9. Hierarchical Structure.

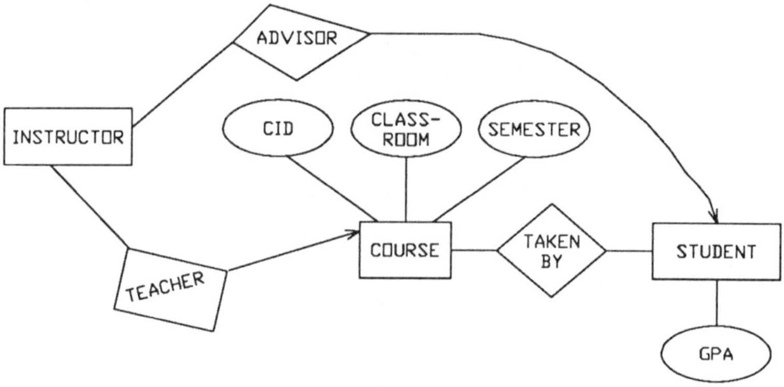

Figure 3-10. ER Model Description.

using our university database, the ER model would provide a structure as shown in Figure 3-10.

To access information, we provide queries of the form:

```
prints students advised by Ting
     and taken by CS 416
```

This data structure supports accessing information based on supplying a function performed on the data sets. In this case the two functions, advised and taken by, are applied to the three data sets to extract the proper information.

Beyond basic data structure and representation, a database does many other things for users. Database management systems perform processing to provide concurrency control update synchronization, crash/failure recovery, integrity checking, and transaction processing. Taken together, these various features provide to users an environment where all users have a view of the database that is consistent across all nodes.

From a LAN perspective, we wish to have a database manager that provides these features, but of the utmost importance to the end user is what he sees—the user interface. The following sections will address this and the basic features of a LAN DBMS.

User Interfaces

User interfaces to database management systems fall into a few categories, procedural and non-procedural.

Procedural interfaces are the ones that we typically see. They are similar to programming languages and require us to produce them in structured ways. Typical of access languages at this level would be the COBOL programming language for hierarchical and network databases; and languages like SQL, QBE, and Quel for relational databases. These require the user to learn a language and follow its means of operation to access the database. More robust user-friendly environments exist, or are under development. These include interfaces to query users by forms, or to provide Icons (pictures) and allow users to access data by pan and zoom primitives. Additionally, natural language interfaces are being examined to allow users query access, by their normal means language. Before these innovative interfaces become widely available, processing power and database processing schemes must improve on their performance. Some examples of database access schemes will be seen in Chapter 6.

Concurrency Control

Concurrency control mechanisms are applied to databases to provide for concurrent, interleaved execution of transactions while preserving the database integrity. This implies that interleaved executions of transactions will not cause the database's state to become incorrect in relation to correctness criteria. The typical means to guarantee correctness is to show that the interleaved executions are equivalent to a strictly sequential execution of the transactions (Figure 3-11). This notion of serial execution is termed serial liability. That is, the interleaved execution is equivalent to the serial execution. This notion of serializability forms the basis for most concurrency control-scheme correctness criteria. The algorithms that have been developed to provide concurrency control deal with solving the read/write and write/write conflict situations. The algorithms fall into three categories: locking, timestamp ordering, and optimistic.

$T_1{}^{j} = R_1(A), R_1(B), W_1(B), W_1(A)$

$T_2{}^{j} = R_2(C), W_2(C), R_2\ B, W_2(B), R_2(A), W_2(A)$

INTERLEAVED EXECUTION: $R_1(A)$, $R_2(C)$, $R_1(B)$, $W_2(C)$, $W_1(B)$, $R_2(B)$, $W_1\ A$, $W_2(B)$, $R_2(A)$, $W_2(B)$

IS EQUIVALENT TO

T_1 FIRST
T_2 SECOND

Figure 3-11. Serializability.

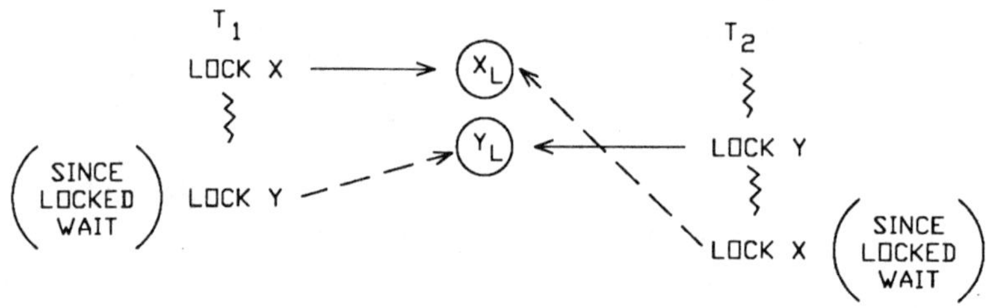

Figure 3-12. Deadlock.

Concurrency control methods using locking require transactions to first acquire all locks they need during their operation; then, as they commit, release the locks to allow them to be available to others. This is referred to as two-phase locking, where the first phase is the growing phase (lock acquisition) and the second phase is the shrinking phase (lock releasing). If all transactions follow the rule that they do not process their transactions until they have acquired the necessary locks and upon completion release them, then there will not be a problem.

This type of concurrency control scheme will require the data manager to keep a list of the database items, which keeps track of locked items and has means to break deadlocking when one transaction has a lock on an item and is looking to acquire another lock on a second item, which is held by another transaction that wants a lock on the first transaction's held item (Figure 3-12). The data manager must decide which transaction is to be aborted (stopped) in order to break the deadlock, and must have a means to restart the failed transaction so the user application can continue.

Locking is one of the schemes found in most database products available today, and does provide good service in low-conflict systems. The problems deal with overhead and how to manage lock tables that may be distributed over the network.

The second form of concurrency control deals with time-stamp ordering. In this scheme, the ordering (sequencing) of transactions and their subparts is performed by the association of a time stamp and the interpretation of this time stamp by the database concurrency control mechanism. To work, time stamps must be unique and allocated in some monotonically increasing or decreasing fashion. The protocol to work requires that each data object record a time stamp for the most recent read (RTS) and a time stamp for the most recent write (WTS). The protocol operates as follows:

1. A transaction T with a time stamp(ts) cannot "Read" an object with a write stamp wts>ts
2. A transaction T with a time stamp ts cannot "write" an object with a read ts rts>ts
3. A transaction T with a time stamp ts can write an object with wts>ts, but the object remains unchanged.

This process will provide a means to guarantee serializability is met as long as a correct means for allocating time stamps exists.

A third form of concurrency control is called optimistic. Techniques of this type allow actions (read/writes) to occur and before they are committed, written to secondary nonvolatile storage, they are certified as correct or incorrect. This certification process requires the checking of all pending transactions against one another and against the latest database activities. The certification checks to see if the applications of the transaction components are serializable, if so accepts the transaction and commits it, else aborts or backs out of the transaction. This process requires the maintenance of extra data to be used in the validation, which can become expensive in some cases.

In all three cases, a LAN adds other problems. We now must be concerned with distributed data and the management of locks, time stamps, or certification tables over distributed sites. These are not simple problems and are still being actively researched by industry and academia.

Update Synchronization

Related to the database issue of concurrency control is that of update synchronization. Update synchronization deals with the LAN problem of updating multiple copies of a data item over a network. Whereas concurrency control deals with the ordering of the reads and writes, update synchronization piggybacks concurrency control write processing. Update synchronization can be performed in real time "immediately" or put off as an afterthought. The time frame is dependent on the user program requirements. That is, if all copies are required to be consistent, then updates must be applied immediately, possibly using locking to lock all copies until updates are performed. On the other hand, if tight consistency is not required, then updates can be performed using other less-costly means. The basis of the following algorithms is that writes are directed by the low-level database components to all sites. Sites operate on them based on their update algorithm and structure. The major bases of update algorithms are the primary copy, two-host resiliency, majority vote, and majority read.

The primary copy update algorithm generates in a master/slave fashion. Updates are directed to a master copy (the primary); this master copy updates its files and issues updates to secondaries based on given heuristics. For example, the primary may update secondaries based on a predefined order or by a timing scheme. In any case, the primary is the only place where the database is guaranteed consistent. Therefore, if users wish to access up-to-date information, they must direct their reads to the primary. In this case, the primary becomes the system bottleneck, and the potential benefits of distributed data are lost.

A second update synchronization scheme is the two-host resiliency scheme. In this algorithm, updates are directed to a master and a secondary at the same time. The two coordinate their updates by a two-phase commit protocol to guarantee that both have consistent databases. The two-phase commit operates as follows:

- The master issues a prepare to commit message to the secondary.
- The master then waits for the secondary to respond ready-to-commit or abort.

- If ready-to-commit is received, the master issues a commit message and then waits for an acknowledgement.
- If an abort is received, the master aborts and indicates its reception of this message.
- Updates are issued to other secondaries at a later time.

In this fashion the two-phase commit provides the two-host resiliency algorithm to have redundant, up-to-date copies of data. Reads can now be directed to either site, thereby increasing concurrency of database operation.

The third scheme for update synchronization in distributed systems is called majority voting. In this scheme, the N nodes comprising the database must vote to determine if an update is to be accepted. The update is issued from the originator to all N sites. The sites in turn decide, based on their present loads, whether to accept this update or reject it. This decision is cast as a vote to the other N-1 sites. To determine if I should ultimately perform the update or not, I must listen and collect N votes from the other N-1 sites. When I have a majority of yes votes, I commit or else abort or roll back the operation.

One can see that intrinsically this process requires much more overhead to accomplish the objective of updating data values. The benefit of this process is that one can read from more sites, potentially increasing concurrency.

The fourth form of update synchronization requires no work up front. That is, one does not have to do any checking with other sites to see if they can perform the update; one simply sends out updates to all, performs it locally, and goes on with business. Likewise, all others perform the same function. Each at their own site determines how to order the updates, possibly on arrival (FIFO), or based on simple node address priority, etc. They perform the updates and go on. The overhead comes on reads. The title "majority read" says it all. When we wish to read an item we query all sites, collect their responses, and decide based on a majority the actual value to use. This scheme favors updates and penalizes reads, whereas the other penalizes updates and favors reads.

Dictionary/Directory

To realize the capability of database systems requires knowledge of the contents thereof. The data dictionary/directory supplies the database manager with its knowledge of data about data ("meta data"). The data stored here can take on many forms. For example the dictionary could provide:

Type of information:

- text, Boolean, character

Structure:

- record
- tree
- etc.

Relationships:

- *a* related to *b*, etc.

Ranges:

- a range from *x* to *z*

Cross references:

- x.y = z.t

Locations:

Statistics:

The sole purpose of a dictionary/directory is to aid the database in making its decisions on processing and allocations. For LANs, it provides the database with addresses to allocate data on remote sites.

Crash Recovery

Due to the nature of databases (the storage and recall of information), they typically in time will experience a failure. Failures and errors occur causing the database to falter and possibly become erroneous. Recovery is the means by which a database is brought back to the point of correct state.

The unit of work and, therefore, recovery is the transaction. The goal in a database system to aid in recovery as well as correct operation is to have all transactions operate in a consistent manner. The transaction either will perform all its intended reads/writes and processing or it will perform none of them. This all or nothing operation of a transaction is referred to as its atomic action. The operation of a transaction is to not allow any of it to occur unless all can occur; but how do we use this notion to keep databases reliable? The answer is that we must keep information about the operations of transactions in order to guarantee this atomic operation in light of possible failures. To do this, databases keep meta data on transactions states in a log. Beyond the log, the database takes snapshots of its state on a periodic basis, referred to as checkpoints. Transactions occurring between checkpoints (Figure 3-13) must have their states

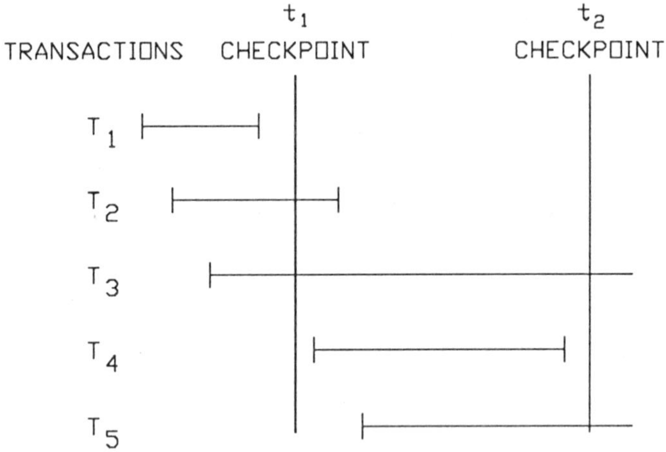

Figure 3-13. Transactions and Checkpoints.

maintained dynamically in the log. This information can be used to recover upon a failure. Recovery is in the form of either/or rollback (remove transaction operations to bring database back to checkpoint), or roll forward (reapply transaction's operations stored in log to bring database up to proper state).

The total recovery process also deals with total transaction failure, where we must abort a failed transaction and restart it at some later time.

The recovery process uses the log to determine, since the last checkpoint (t_1 in Figure 3-13), what transactions must be redone and those to undo. It must redo all completed transactions, such as T_2 and T_4 in Figure 3-13. It must abort all transactions in an unknown state, such as T_3 and T_5 in Figure 3-13, undo their changes on the database, and restart them in the future. To accomplish this, the log must maintain all database changes in chronological order, as well as before images of database items for undo's, and database items after images for redo's.

Logs in databases with recovery use what is referred to as write-ahead log. This form writes to the log first before incorporating the changes on the physical database. The log must be stored in nonvolatile storage for the recovery mechanisms to work. The use of logs and checkpoints relieves the database from maintaining a total log of all actions, we only require storage of action between checkpoints. Checkpoints become the boundaries of recovery. For example, in Figure 3-13, if T_2 is a failure mark versus a checkpoint, the following would need to be done with the five transactions shown.

Transaction T_1, since it was incorporated (committed) before checkpoint t_1, needs to have nothing done with it. T_2 and T_4 have committed (at least to the log) before the failure; therefore, they must be redone. Redone implies that all actions performed by

T_2 and T_4, since t_1 up to the failure at t_2, must be reapplied, beginning with the status at checkpoint t_1 until completion. Likewise, T_3 and T_4, since they are not completed (committed) by the time of the failure, must be aborted (stopped) and rolled back (undone). This is performed simply by applying the before image of their database components and restarting their transactions from that point.

T_2 and T_4 can be recovered more quickly by simply applying them after image (stored in log) versus recomputing their entire log of operations. One can see that without recovery a database can and will ultimately fail, causing loss of stored information. Due to this problem, any database purchased should have some form of logging and checkpointing available to enable recovery.

In a LAN environment, recovery mechanisms need an additional piece—they require two-phase protocols to synchronize the operations of the checkpointing and counting processes.

Integrity Checking (Consistency)

Integrity of the database refers to its accuracy, correctness, or validity. Integrity addresses the problem of guarding the information against invalid or erroneous updates. To provide integrity, the database must store meta data that classifies the rules of integrity to be applied to database actions. In some cases, these rules can account for up to 80% of a database specification. The integrity checker using these rules must monitor transaction updates and detect when erroneous values (breaking of rules) are being applied. It must then determine how to act on these updates; i.e., it must report problems, reject update if critical, and possibly correct update values if possible.

The integrity subsystem uses the rules defined to determine what to check and how to react on encountering a violation. Integrity checkers have three components: a trigger (when to react), a constraint (what to look for), and a response (based on trigger and constraint).

Typical integrity rules deal with domain checking (boundary conditions) and structure. To use this feature of a database one must be willing to suffer the added overhead that this can cause.

Security

As important as total system security is database security. Security is needed to keep unwanted elements from using information or changing/destroying information. The aspects of security are physical and logical security.

Physical security deals with access to facility. That is, one must not allow unwanted users physical entry to the building where the database is stored; nor do we want them to have access to the room or office where the device is. To provide physical security requires activities such as guards or key locks, camera monitoring systems, and other physical constraints. Additionally, to guard against tampering, we must provide adequate maintenance and backup on data repositories to maintain their level of deterrence.

Logical security on the database deals with issues of authorization. Authorization can be multileveled in the sense that we can control it from various levels of granularity. We can check authorization on initial entry to the database function, or at each underlying level—i.e., database, files, relations, tuples, objects. In each case, the level of security checking drives the overhead in providing it. That is, the lower the granularity of security authorization, the higher the cost in overhead one must pay to provide it.

Typical schemes to provide security include passwords, access matrices, and access lists. These provide a means to check a user's authorization to be on the system, as well as rights to individual items. Details of these modes of security can be found in [Fortier 1986] and other books on operating systems.

Query Processing and Optimization

As databases have progressed and the volume of data that is processed increases, the need for better processing solutions has arisen. Query processing deals with the determination of how to access information, given a user request. In the worst case, the query processor simply gets the data sequentially and processes them based on how the query is written (i.e., they follow the order given in the query). On the other hand, if this approach is taken, query processing of simple form could take enormous amounts of time, as was shown in an example in Date [1986] that indicated good processing strategies can reduce query-processing time from days to seconds.

The scheme to provide this better way is query optimization. These are algorithms designed to use the meta data stored in dictionaries/directories to determine how this query can be processed in the fewest steps possible and at a reduced cost. As an example, the case of a simple join of two tables with selection qualifiers is shown below.

```
Select sname
 from students,faculty
 where student.GPA>3.0
 and faculty name=Jones
 and faculty#=advisor#
```

This query looks to select a list of students who have Jones as an advisor and have a grade point average greater than 3.0. The simple graph structure of this query is shown in Figure 3-14. In this figure we show the worst case operation of this query, where faculty and student relations are joined as a Cartesian product creating a relation of size sXf. This joined relation is then scanned one at a time for tuples matching the given qualities listed above. This query could be optimized easily by forcing the

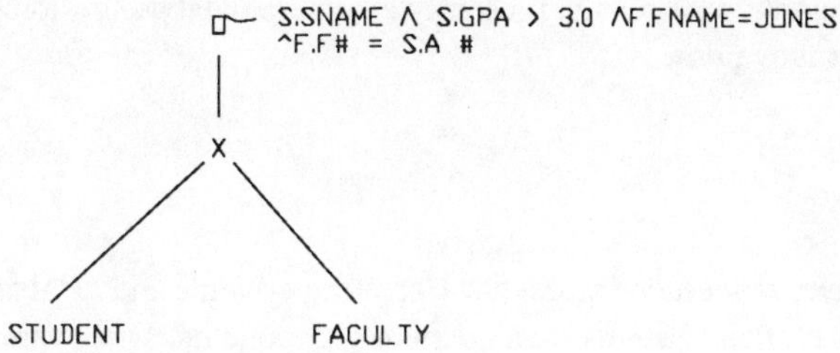

Figure 3-14. Graph Structure of Query.

selections with qualifiers down to the lowest leafs, and performing an equijoin on s.advisor # and F.Faculty# shown in Figure 3-15.

This reduced query operates by determining only the faculty whose last name is Jones, then projecting just their faculty numbers. On the other branch the student relation is scanned to select only the tuples with Gpa>3.0, then projecting only the A# and Sname from these tuples. These are then joined on F#=A# to provide the names of the students who meet the criteria of having one of the Jones as an advisor. In this fashion, only the information needing the query is passed up from leg to leg.

This will shrink the volume of data that needs to be processed at each step, thereby reducing the overall time to perform this function.

In a LAN environment, distributed databases exist and exacerbate the problems of query processing. Although without some intelligence built in, the cost of having a database may be too great to be useful, query processing in some form is paramount in a LAN environment.

The features described in this section are found in most database systems in minicomputers and mainframe environments. Future LAN environments with PC workstations, as well as larger machines, will develop and use these same features.

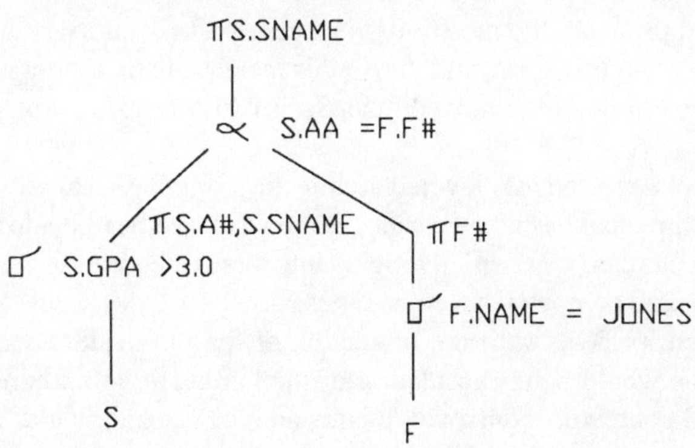

Figure 3-15. Optimized Query.

Chapter 8 will review some of the presently available database systems and indicate the qualities that they possess.

SECURITY

Systems management includes more than operating systems and database software. It also encompasses other systems management components. One typically taken for granted and often overlooked is security.

Security deals with the protection of resources against unauthorized disclosure, modification, restriction, or destruction. In the early days before networks and before computers, we only needed a good security guard, a safe, and a shredder to delete information when it didn't need to be stored. Computers added a new dimension. It was not just the guard and a safe room, but also controls against access once into the room.

In the old days this was accomplished by a lock on the safe. In the computer system the same holds, although instead of a lock on a safe, it is the password on the computer. The password, like the combination lock, lets you into the safe. Once in, though, unless the safe was compartmentalized with additional locks, you could read anything. Likewise, the early computers provided means to restrict getting in, but once in anything goes.

As computers went from single machines to multiple machines and then to networks, the problems got worse. Communications lines became like a safe with a big hole in it. Anyone could get in by tapping the weakness and using it to acquire information. A network becomes the loose link in a security blanket, since it can in many cases be easily tapped, providing an intruder with the ability to listen to systems traffic and possibly to enter the system.

Data security is needed more now since the number of computers has gone up, as has the number of computers linked via wide area networks and LANs. More people are versed in the design and use of computers and, therefore, we must make computers more secure from these new "intelligent hackers."

And last and probably the most important reason for security is that more data is being stored in computer form; and due to this increase in data storage and use, more valuable data worth stealing is now online. Security is needed to protect this information.

Security can be viewed as a layered architecture (Figure 3-16) with many components. Each component has its own aspect as to what overhead it adds to system use. These components can be classified roughly into three categories of security: physical, administration, and computer system and networks.

Physical security deals with the protection of the physical assets of the system. Included in this would be encapsulation of the hardware into secure environments (e.g., protected computer rooms with guards and locks, shielding of network cables to make eavesdropping more difficult).

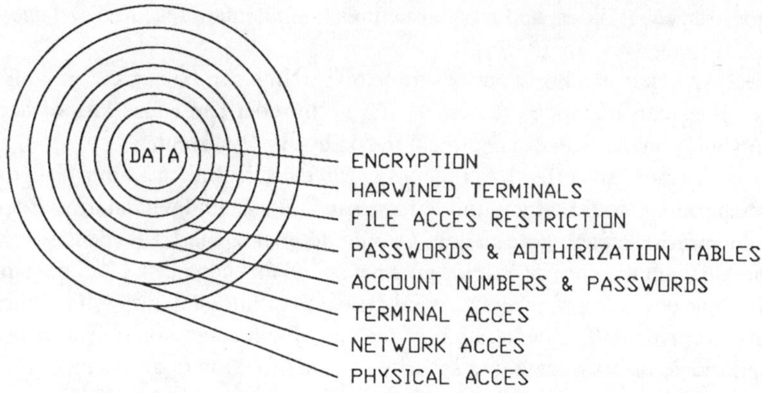

Figure 3-16. Layers of Security.

Administrative security deals with the procedures in place for allowing access (clearing people), collecting, validating, processing, controlling, and distributing data. This aspect decides how to perform security. For example, what controls do we put on people (badges, key-code access, guard sign-ins, compartmental machines, terminals, etc)?

The last component is the physical computer system and the network. This level of security deals with the nuts and bolts of security—for example, user identification and authentification, encryption of data, passwords, etc.

User authentification is the system's means to verify that a user is indeed who he says he is. For example, this could be accomplished by visual interrogation, fingerprints, retina patterns, voice patterns, or via confidential knowledge such as account numbers, passwords, hidden test programs, and patterns.

In the LAN environment, we must determine that the user is authorized for net access once they have been cleared. In a LAN environment, one must be careful that all users, remote or otherwise, are checked as to their authenticity and rights of use. Another form of authentification deals with data.

Data authentification refers to methods to ensure that the data flowing in the system and being infused into the system is legitimate data, not erroneous or possibly even viral data. Authentification must be able to determine data sources and validate their accuracy and validity as a source, thereby freezing out potential and counterfeit sources. Access control to the system is typically managed by a simple login procedure where the user is required to provide a username and password for general entry, and possibly other passwords for entry into protected files and programs, etc.

Resource access control deals with controlling access to underlying network resources. For example, if a user wishes to log on remotely to another machine, the network must check if this user from this site is authorized for such actions. Typical mechanisms for providing this type of security include access control matrices, access control lists, capabilities, and other structures that maintain resource and user access rights.

The LAN itself is a big security problem. The line can be cut, causing all to lose access; lines can be tapped (ethernet) and an intruder can cause loss of access by putting noise on the line, or acquire all traffic by simply listening.

To rectify these and other LAN security problems, the link itself must be protected. For example, we could bury it underground and put a shield with alarms around it to provide security breach notification. One could encase the LAN media in conduits that are difficult to enter, or provide an inner shield that can mask LAN transmissions from simple eavesdropping and provide detection of intrusion through monitoring of signals over shield. The penetration of the shield will cause some disturbance in its transmissions, thereby providing security with notification of an attempted break in. A more viable solution for the LAN media is to mask the data being sent, using encryption techniques.

CRYPTOGRAPHY

Cryptography has been in use for thousands of years as a means of transferring information from one location to another without making the information public and preventing interception by unwanted groups.

Basic types of ciphers will be discussed in the following sections along with some cryptoanalytic techniques for breaking them.

Ciphers

Substitution ciphers date far back in history and, with modifications, are the basis for some of today's ciphers. A simple example of a substitution cipher is the Caesar Cipher that was used by Julius Caesar.

First, write the alphabet followed by the figures 0 through 9. Below this, write the same sequence cyclically shifting to the right any arbitrary number of units, as shown in Figure 3-17. Take each symbol in the plain text, locate it in the top row, and replace it with the symbol in the row below. An "A" would become a "0" and so on. This type of cipher is commonly called a "Shift Cipher." In this example, the shift count or key is 10.

The simple shift cipher can be extended into an unbreakable cipher called the "One-Time Pad." One-Time Pads are used on the Hot Line between Moscow and Washington. The keys are exchanged periodically through the embassies.

The One-Time Pad can be created by using the device shown in Figure 3-18. As each plain-text character is encoded, the dial is spun, revealing the substitution character and the shift count. If a "T" is going to be encoded and an "F" is selected by the dial, this is the same as a shift of 14 in the bottom row of Figure 3-17. The "F" is the cipher text and the shift count (14) is the key. Each character of the plain text is

A	B	C	D	E	F	G	H	I	J	K	L	M	N	O	P	Q	R
0	1	2	3	4	5	6	7	8	9	A	B	C	D	E	F	G	H
S	T	U	V	W	X	Y	Z	0	1	2	3	4	5	6	7	8	9
I	J	K	L	M	N	O	P	Q	R	S	T	U	V	W	X	Y	Z

Figure 3-17. A Caesar Cipher with a 10-Shift.

enciphered this way. The key is a list of random numbers and must be exchanged and used by both the encoding and decoding parties.

The key is never used more than once. The key must be the same size as the message and is destroyed immediately after its use.

Although this technique is secure, the exchange of cipher keys that are of equal length with the message is a significant problem with large messages. The disposal of the large number of keys is also a significant problem. These difficulties have limited the use of One-Time Pads to military application.

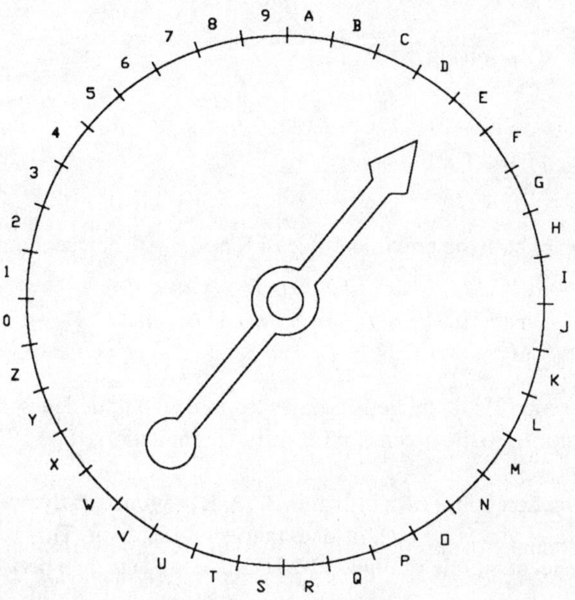

Figure 3-18. Randomizer for Encoding a "One-Time Pad."

The National Bureau of Standards Data Encryption Standard (DES) is a data encryption, decryption standard developed for private industry. The DES has been implemented by several companies on LSI chips. The standard was initially developed for IBM and was modified for NBS.

The DES enciphers and deciphers 64-bit blocks of binary data under control of a 64-bit key. This is different than the substitution and One-Time Pad in that they work on a character basis. The DES is intended to work with computers and computer-generated data.

Enciphering and deciphering are performed with the same 64-bit key. The enciphering and deciphering processes are the reverse of one another. This is accomplished by revising the order of scheduling with which the key-dependent processing is performed. A block of data is subjected to an Initial Permutation (IP), then to a set of key-dependent computations, and then to the inverse permutation IP-1. The key-dependent computation can be defined in terms of a function of a Key Schedule (KS). The 64-bit block (LR) is divided into two equal blocks denoted as "L" for left and "R" for right.

The enciphering process is depicted in Figure 3-19. The 64-bit input data LR is subjected to the initial permutation shown in Figure 3-20. Bit 58 of LR is the first bit of the permuted output, but 50 is second, and so on, with bit 7 being the last output bit.

The output of IP is then subjected to 16 iterations of a cipher function "F," which operates on L and R under control of the output of the Key Schedule (KS). The state is described by the following equation:

$$L_n = R_{N-1} \quad (1)$$
$$R_n = L_{N-1} + R(R_{n-1}, K_n) \quad (2)$$

where

$$n = 1, 2, \ldots, 16$$

F is a function of both the previous value of R and K_n ,the present output of KS, the Key Schedule.

$$K_n = KS\ (n, Key) \quad (3)$$

Equations (1) and (2) are iterated 16 times to yield an output denoted $R_{16}\,L_{16}$. This output is then applied to the inverse of the initial permutation (IP–1). This is depicted in Figure 3-21.

The cipher function F is shown in Figure 3-22. E is a function that takes R_n , which is 32 bits long, and yields a block of data that is 48 bits long. This is accomplished with the E-bit selection table in Figure 3-23. Equation (4) is then performed.

$$E(R_n) + K_n \quad (4)$$

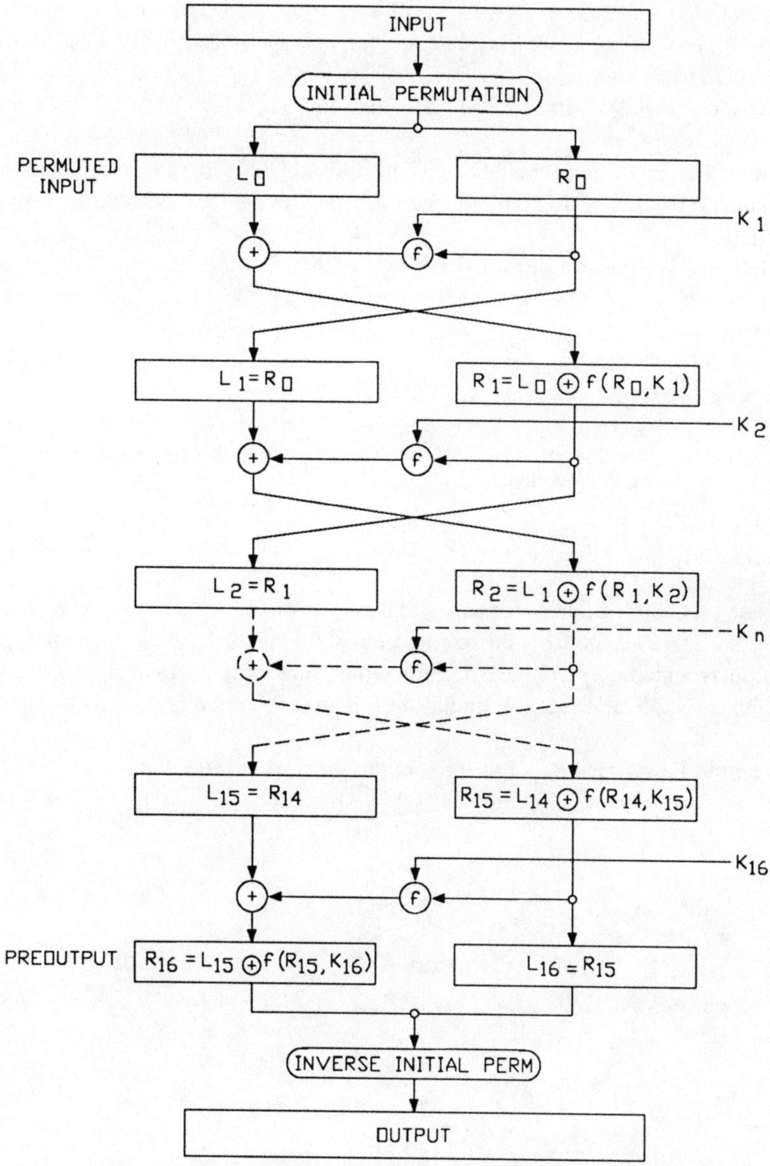

Figure 3-19. Enciphering Computation.

The output of E and the key schedule is then applied to the selection function S_1 (Figure 3-24). The 48-bit input to S is grouped into 8 six-bit blocks. One block is applied to each selection function. Each six-bit block is grouped into a two-bit row

IP

58	50	42	34	26	18	10	2
60	52	44	36	28	20	12	4
62	54	46	38	30	22	14	6
64	56	48	40	32	24	16	8
57	49	41	33	25	17	9	
59	51	43	35	27	19	11	3
61	53	45	37	29	21	13	5
63	55	47	39	31	23	15	7

Figure 3-20. Initial Permutation.

and a four-bit column address for use with the S lookup tables. For example, if one of the blocks is 011011, the 01 is the row address and 1101 or 13 is the column address. The output from table S_1 would be 5. The s tables transform the 48 computer bits back to 32 bits for application to a permutation function P, which is described in Figure 3-25.

The output P becomes $R_{n = 1}$ or the input to the next iteration of F.

IP-1

40	8	48	16	56	24	64	32
39	7	47	15	55	23	63	31
38	6	46	14	54	22	62	30
37	5	45	13	53	21	61	29
36	4	44	12	52	20	60	28
35	3	43	11	51	19	59	27
34	2	42	10	50	18	58	26
33	1	41	9	49	17	57	25

Figure 3-21. Inverse of Initial Permutation.

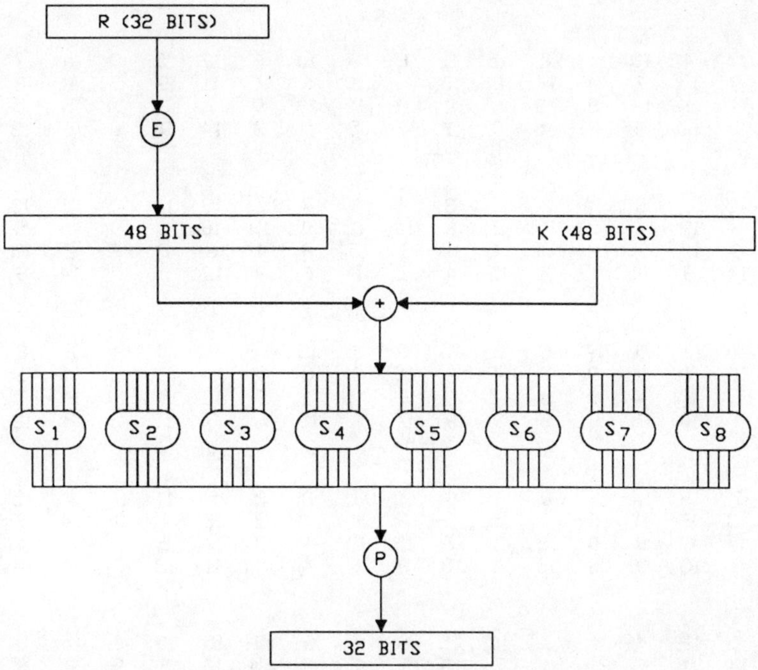

Figure 3-22. Calculation of f(R,K).

E BIT-SELECTION TABLE

32	1	2	3	4	5
4	5	6	7	8	9
8	9	10	11	12	13
12	13	14	15	16	17
16	17	18	19	20	21
20	21	22	23	24	25
24	25	26	27	28	29
28	29	30	31	32	1

Figure 3-23. E-bit Selection Table.

S_1

14	4	13	1	2	15	11	8	3	10	6	12	5	9	0	7
0	15	7	4	14	2	13	1	10	6	12	11	9	5	3	8
4	1	14	8	13	6	2	11	15	12	9	7	3	10	5	0
15	12	8	2	4	9	1	7	5	11	3	14	10	0	6	13

S_2

15	1	8	14	6	11	3	4	9	7	2	13	12	0	5	10
3	13	4	7	15	2	8	14	12	0	12	10	6	9	1	5
0	14	7	11	10	4	13	1	5	8	1	6	9	3	2	15
14	8	10	1	3	15	4	2	11	6	7	12	0	5	14	9

S_3

10	0	9	14	6	3	15	5	1	13	12	7	11	4	2	8
13	7	0	9	3	4	6	10	2	8	5	14	12	11	15	1
13	6	4	9	8	15	3	0	11	1	2	12	5	10	14	7
1	10	13	0	6	9	8	7	4	15	14	3	11	5	2	12

S_4

7	13	14	3	0	6	9	10	1	2	8	5	11	12	4	15
13	8	11	5	6	15	0	3	4	7	2	12	1	10	14	9
10	6	9	0	12	11	7	13	15	1	3	14	5	2	8	4
3	15	0	6	10	1	13	8	9	4	5	11	12	7	2	14

S_5

2	12	4	1	7	10	11	6	8	5	3	15	13	0	14	9
14	11	2	12	4	7	13	1	5	0	15	10	3	9	8	6
4	2	1	11	10	13	7	8	15	9	12	5	6	3	0	14
11	8	12	7	1	14	2	13	6	15	0	9	10	4	5	3

S_6

12	1	10	15	9	2	6	8	0	13	3	4	14	7	5	11
10	15	4	2	7	12	9	5	6	1	13	14	0	11	3	8
9	14	15	5	2	8	12	3	7	0	4	10	1	13	11	6
4	3	2	12	9	5	15	10	11	14	1	7	6	0	8	13

S_7

4	11	2	14	15	0	8	13	3	12	9	7	5	10	6	1
13	0	11	7	4	9	1	10	14	3	5	12	2	15	8	6
1	4	11	13	12	3	7	14	10	15	6	8	0	5	9	2
6	11	13	8	1	4	10	7	9	5	0	15	14	2	3	12

S_8

13	2	8	4	6	15	11	1	10	9	3	14	5	0	12	7
1	15	13	8	10	3	7	4	12	5	6	11	0	14	9	2
7	11	4	1	9	12	14	2	0	6	10	13	15	3	5	8
2	1	14	7	4	10	8	13	15	12	9	0	3	5	6	11

Figure 3-24. Functions of S_1 through S_8.

The Key Schedule is used to calculate K_n for use in Equations (2) and (4). The Key Schedule calculation is shown in Figure 3-26. Fifty-six of the original 64 bits of the key are used by the Key Schedule calculation. The eight stripped-off bits are parity bits used for error detection. The 56 bits of the key are permuted by means of the table

P

16	7	20	21
29	12	28	17
1	15	23	26
5	18	31	10
2	8	24	14
32	27	3	9
19	13	30	6
22	11	4	25

Figure 3-25. Permutation Function P.

in Figure 3-27. The top four rows determine C_o and the bottom four rows determine D_o . Bit 1 of the key will become bit 50 of C_{vov} and bit 63 of D_o . C_o and D_o are then shifted according to the shift table in Figure 3-27. The shifted outputs are passed on to the next state and also applied to Permuted Choice 2 before being output as K_n .

Permuted choice 2 is shown in Figure 3-28 and is similar to permuted choice 1 in principle.

Deciphering is accomplished by reversing the procedure described previously. Figure 3-29 depicts the deciphering process. The 64-bit cipher text is applied to IP^{-1} and then through stages 16 through 1 of F. L_o and R_o are then applied to IP and then to the output.

Public-Key encryption was first suggested by Diffie and Hellman in 1975. Public-Key cryptography uses a class of functions that are "unbreakable." In principle, these ciphers can be broken, although only at great computer time and expense. Public-Key encryption has two keys. One is published and the other is kept a secret. The Public-Key is used to encrypt data and the secret key is used to decrypt the data. If individual A wishes to send a message to individual B, he looks up B's public key and encrypts the message. B receives the message and decrypts the message using his secret key. This is depicted in Figure 3-30. If the message is intercepted by C, the message cannot be deciphered because C does not know B's secret key.

The encryption algorithm is also public knowledge, but is such that data can be enciphered easily but cannot be deciphered using the public key. Such a function is called a one-way function. A simple example of a one way function is

$$f(X) = X^4 + 5X^3 + X + 10 \quad (5)$$

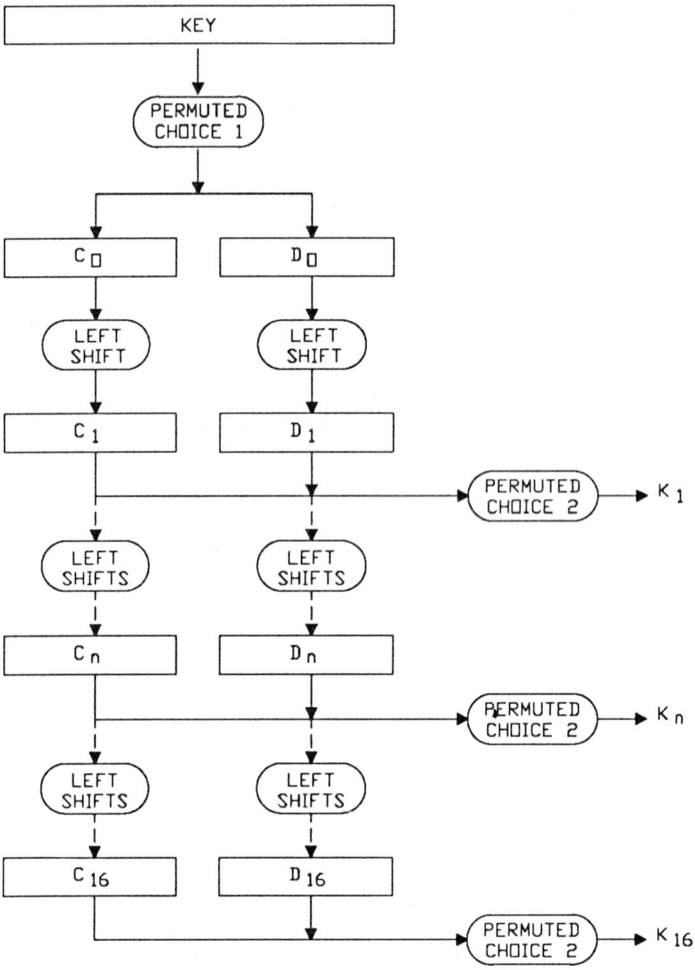

Figure 3-26. Key Schedule.

If X = 10 is the plain text information, f(X) = 15020 is the cipher text. The calculation of f(X) is very straightforward. Given Equation (5) and f(X) = 15020, the value of X = 10 is not as easy to calculate. The type of one-way function often used in Public-Key cryptography is called a trapdoor function. A trapdoor function is a one-way function except that a secret means has been built into the function to allow easy deciphering. This information is very carefully hidden and without the secret key the cipher text cannot be deciphered.

Iteration Number	Number of Left shifts
1	1
2	1
3	2
4	2
5	2
6	2
7	2
8	2
9	1
10	2
11	2
12	2
13	2
14	2
15	2
16	1

PC-1

C_0	57	49	41	33	25	17	9
	1	58	50	42	34	26	18
	10	2	59	51	43	35	27
	19	11	3	60	52	44	36
D_0	63	55	47	39	31	23	15
	7	62	54	46	38	30	22
	14	6	61	53	45	37	29
	21	13	5	28	20	12	4

Figure 3-27. Permutation Table.

The Public-Key technique also allows itself to be used for a signature. Individual C, in the previous example, would have the means to send messages to B by the same method as A. B would have no idea which individual has really sent the message. Figure 3-31 shows how a digital signature can be sent by Public-Key cryptography. User A takes his name and encrypts it using his secret key D_A. He then further encrypts

PC-2

14	17	11	24	1	5
3	28	15	6	21	10
23	19	12	4	26	8
16	7	27	20	13	2
41	52	31	37	47	55
30	10	51	45	33	48
44	49	39	56	34	53
46	42	50	36	29	32

Figure 3-28. Permuted Choice 2.

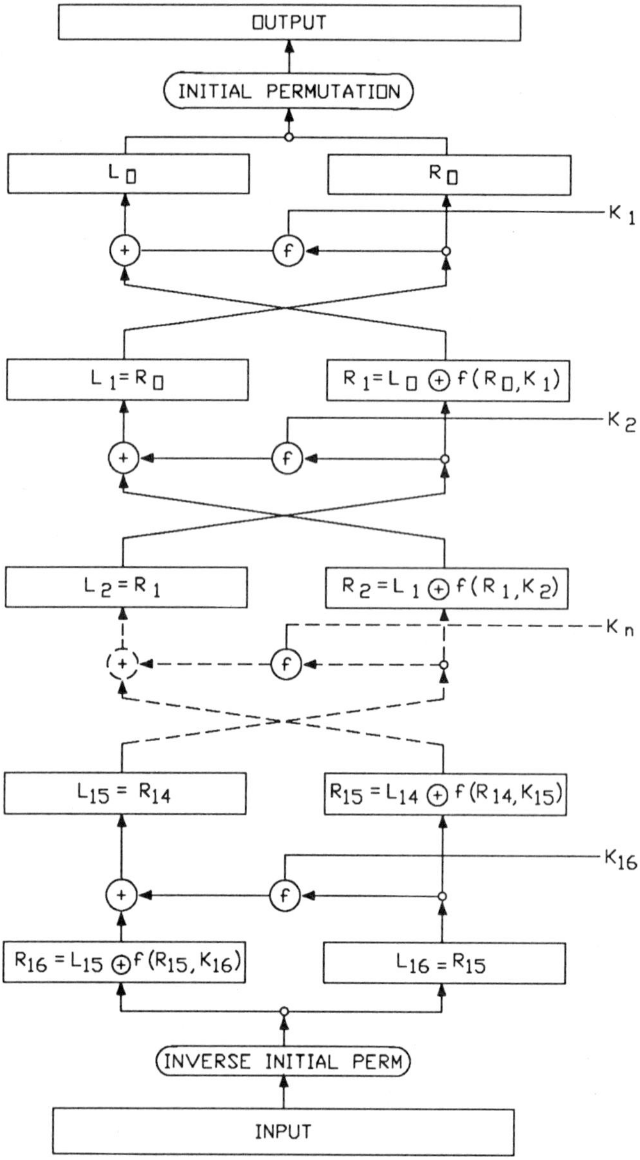

Figure 3-29. Deciphering Computation.

it by using B's public key E_B. B deciphers the message by first using his own secret key D_B and then applies A's public key E_A. The result is the original message (A's name). B is the only possible sender of the message because he is the only one who knows his own secret key.

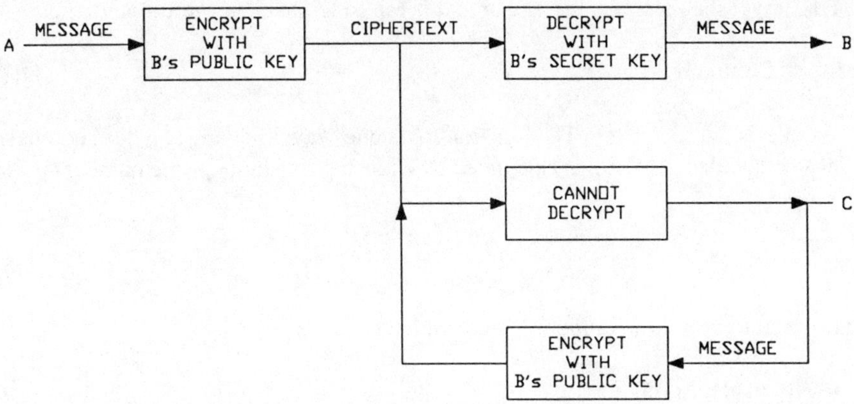

Figure 3-30. Public-Key Functional Diagram.

The advantage of Public-Key crypto systems are many: they are secure, and they have eliminated the need to transfer the key by a special courier before two individuals can begin secure communications.

The following summarizes a Public-Key cryptosystem:

a. Deciphering (D) the enciphered (E) form of a message (M) yields M.

$$D[E(M)] = M$$

b. Both E and D are easy to compute.
c. By publicly revealing E, the user does not reveal an easy way to compute D.
d. If a message M is deciphered and then enciphered, M is the result:

$$E[D(M)] = M$$

Rivest at MIT was the first to propose a viable Public-Key cryptosystem. This system uses a trapdoor function and works as follows: to encrypt a message, the message M is raised to the eth power, modulo n; the e,n is the Public Key.

$$C = M^e \bmod n \tag{6}$$

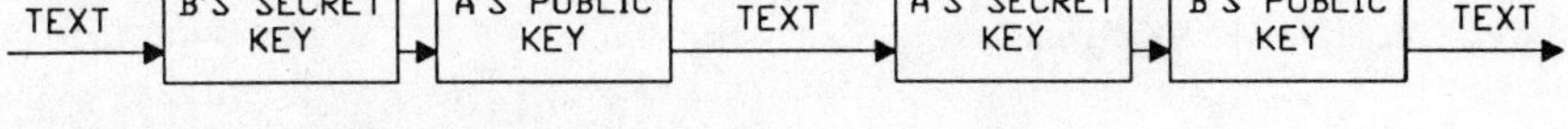

Figure 3-31. Digital Signature in Public-Key Cryptography.

To decrypt the message, the cipher text is raised to the dth power, modulo n.

$$M = C^d \bmod n \tag{7}$$

The d, n is the secret key. The function (6) is one-way; knowing e, n does not make it easy to calculate d. The n is calculated by choosing two large prime numbers p and q.

$$n = p^*q \tag{8}$$

d is chosen to be a large random integer which satisfies

$$\gcd [d, (p-1)^*(q-1)] = 1 \tag{9}$$

e is calculated from d,p,q in such a way that it is the multiplicative inverse of d

$$(e^*d) \bmod (p-1) * (q-1) = 1 \tag{10}$$

If n could be factored into p and q, the code would be broken. With present numerical techniques, this is an enormous problem. This is the one-way function that makes this technique secure. Figure 3-32 shows key size vs. time-to-break-key in operations. A numerical example of the Rivest technique for encryption and for a signature transmittal is given at the end of this chapter.

Hillman and Merkle described another Public-Key crypto system using trapdoor knapsacks. This technique uses weighted n dimensional vectors to encode binary information.

$$C = M_i^*e_i \tag{11}$$

Digits	Number of Operations	Time
50	1.4×10^{10}	3.9 hours
75	9.0×10^{12}	104 days
100	2.3×10^{15}	74 years
200	1.2×10^{23}	3.8×10^{9} years
300	1.5×10^{29}	4.9×10^{15} years
500	1.3×10^{39}	4.2×10^{25} years

Figure 3-32. Number of Operations Needed to Factor n.

The designer selects a knapsack vector e_i' such that (the next weight is

$$e_i > \sum_{j=1}^{i-1} e_j \qquad (12)$$

larger than the sum of all weights preceding it, and $C' = e' \times M$ is easily solved.

The designer can compute

$$C' = w^{-1} * C \bmod m \qquad (14)$$

$$= w^{-1} M_i * i_i \bmod m \qquad (15)$$

$$= w^{-1} M_i \times w \times e_i \bmod m \qquad (16)$$

$$= M_i * e_i' \bmod m \qquad (17)$$

If m is chosen so that m e_i then (17) implies that C′ is equal to $M_i * e_i'$ in integer arithmetic as well as mod m. This can easily be solved for M. Appendix A gives a numerical example of the trapdoor knapsack technique. The e is the encryption key and m, w, and e′ are the secret keys. To be secure, keys should be of the size 2^{100}. This will create an expansion in the number of bits of 200 to 1.

Cryptoanalysis

There are three basic ways of attacking the cipher. These techniques are based on the type of information available to the cryptoanalyst. The attacks are defined as follows:

1. Ciphertext attack—Only encrypted data available.
2. Plaintext attack—Plaintext and ciphertext are known for a few pieces of data.
3. Selected Plaintext attack—Cryptoanalyst has access to encryption device and can process an unlimited number of known messages.

It is obvious that substitution ciphers such as the Caesar cipher are broken if plaintext and ciphertext are known for the same key. One simple technique for breaking the Caesar cipher would be to scan resulting characters for their frequency of occurrence. It is a well-known fact that the E occurs 13% of the time in the English language. Once one character is known in a simple shift cipher, the key has been found. More complicated substitution may require analysis based on word lengths, repetitiveness of messages, and symmetry in the output. Substitution ciphers can generally be broken under a concerted ciphertext attack.

The One-Time Pad is an exception to the rule. Since the key (shift count) is changed after every character in a random manner, the code is unbreakable. At any point in the message, one character has the equal probability of being any other character. The key is destroyed after use, so any possible message to message correlation is not possible. The One-Time Pad cannot be broken by any of the known types of attack.

The DES is embroiled in much controversy over the degree of its security. It is generally agreed that the DES must be secure to a plaintext attack. Diffie and Hellman maintain that the DES is only secure for the next 5 to 10 years before it will be economically feasible to break the code. Diffie and Hellman maintain that a machine could be built (projecting technology and costs ahead 10 years) for $200,000 and a solution in less than 10 hours for $50. The machine is based around the DES chip itself and the use of a plaintext attack. The scheme used is brute force and is called an exhaustive search. The machine would be constructed around 10^6 DES chips, all searching for the correct key in parallel. Many people including the NBS have questioned these figures on cost, maintaining that such a machine would be impossible and too costly to produce within the next 10 years. Further controversy exists over the possible existence of a trap door. Hellman has shown that it is possible to create a trap door with DES by manipulation of the various permutations and lookups that occur in the processing. No one has found a trap door in DES to date. A trap door would greatly reduce the time and cost needed to find a particular key. The DES was studied for 17 man-years for the existence of a trap door. The question exists that if they had found a trap door, would they have told anyone? A Congressional investigation was mounted and determined that DES is secure. Hellman maintains that the 56-bit key length should be longer, maybe 128 bits. Adding 72 more bits would increase the number of operations in an exhaustive search 2^{72} times, making the code virtually impossible to break within 10 years or even longer. The argument now centers around how this can be done. The NBS believes that this can be accomplished by connecting two DES chips in series, with each using a different key, effectively doubling the key length. Hellman believes and intends to prove that this will not work, but will have an effect that can be shown in a shift cipher. If a shift cipher of 10 were enciphered again with a shift of 5, the complexity has not been increased—a shift cipher of 15 has been created. Although this is a simple example, connecting two DES chips might yield less than a 2 x 56 bit key. Hellman maintains that the basic DES should be changed to a larger key before it has proliferated into a large number of equipments.

Public-Key cryptosystems have only been tested by exhaustive search techniques. The Public-Key cryptosystem that raises the message M to a power could be broken if a better factorization algorithm were available. The numbers in Figure 3-32 are based on the fastest known algorithm for factorization. The Public-Key cryptosystem is based on the difficulty of certain types of calculation for its security. More information is available for the cryptoanalyst to use in breaking the cipher than with the DES. As mathematical theory advances, even these techniques may be broken.

Summary of DES and Public-Key Cryptosystems

DES and Public-Key crypotsystems are both commercially feasible with today's technology. Public-Key, as discussed in this chapter, has many good points over DES and a few bad points.

DES requires the keys be transferred by some secure means, probably via courier. This can be expensive and inconvenient. Public-Key is extremely convenient in this regard. Public keys can be published in books or stored in write-protected databases.

DES is computationally very attractive and the algorithm is fast and can be easily built on a single chip. Public-Key cryptosystem, as described, is very computation-intensive, as in the MIT system. High throughput rates are unlikely. The knapsack technique has a large data expansion (200:1)—this would cut down throughput on a communication channel.

A compromise could be reached by using Public-Key systems for use as secure channels of transferring DES-type keys. The keys are short and would not require much time to encrypt using Public-Key algorithms. The inconvenience of the courier could be eliminated, and the DES would be made more secure with more frequent changing of keys possible and economical. The signature aspect of the Public-Key would guarantee that you were exchanging keys with the correct individual.

Example of Public-Key Cryptosystem

USER A

$p = 2$ $q = 5$ (arbitrary choice)
$n_A = p * q = 10$
$d_A = 7$ (arbitrary choice)
$e_A = 3$ from $e * d = 1 \pmod{(p-1 * q-1)}$
Public Key = 10, 3 = (n,e)
Secret Key = 10, 7 = (n,d)

USER B

$p = 2$ $q = 7$ (arbitrary choice)
$n_B = 14 = p * q$
$d_B = 11$ (arbitrary choice)
$e_B = 2$ from $e * d = 1 \pmod{(p-1 * q-1)}$
Public Key = 14, 2
Secret Key = 14, 11

To send a message = 2 from A to B then $M^e \bmod n = C$

$$C = M^{e}B \bmod n_B = 2^2 \bmod 14$$
$$C = 4 \bmod 14 = 4$$

To decrypt

$$C = M^{a}B \bmod n_B = 4^{11} \bmod 14$$
$$= 4194304 \bmod 14 = 2$$

To send a signature from A to B, first decrypt using A's secret key.

$$C_1\ M^{d}A \bmod n_a = 2^7 \bmod 10 = 128 \bmod 10 = 8$$

then encrypt using B's public key.

$$C = M_{1Pv}{}^{e}B \bmod n_B\ 8^{\ 2} \bmod 14 = 64 \bmod 14$$
$$= 8$$

To decrypt the signature, first use B's secret key.

$$C' = C^{d}B\mathrm{mod}\ n_B = 8^{11} \bmod 14 = 8589934592 \bmod 14$$
$$C' = 8$$

Then use A's public key

$$M = C'^{e}A \bmod n_A = 8^3 \bmod 10 = 512 \bmod 10$$
$$M = 2$$

If A had not sent the message, then use of A's public key would have resulted in the improper signature after decrypting.

SUMMARY

This chapter introduced three of the major components of system management, namely:

- Operating System
- Database Manager
- Security

The goal of the chapter was to introduce readers to the construction of each component and the various technologies utilized in secure environments.

CRYPTOGRAPHY REFERENCES

Data Encryption Standard, FIPS Publication 46, National Bureau of Standards.

Diffie, Whitfield, and Martin E. Hellman. "New Directions in Cryptography." In *IEEE Transactions on Information Theory*, November 1976.

Rivest, R. L., Shamir, A., and Adelman, L. "A Method for Obtaining Digital Signatures and Public Key Cryptosystems." In *Communications of the ACM*, February 1978.

Merkle, Ralph C., and Hellman, Martin E. "Hiding Information and Signatures in Trapdoor Knapsacks." In *IEEE Transactions on Information Theory*, September 1978.

Diffie, Whitfield, and Hellman, Martin E. "Exhaustive Cryptoanalysis of the NBS Data Encryption Standard." In *Computer*, 1977.

Sugarman, Robert. "On Foiling Computer Crime." In *IEEE Spectrum*, July, 1979.

4

TRANSPORT/MONITORING

INTRODUCTION

The transport/monitoring component of a LAN hierarchy provides for "correct and consistent" service over the LAN. This collection of services provides for the error-free transmission of information. It performs the jobs of monitoring the LAN activities, detecting error conditions, correcting them if possible, and if not providing services to reconfigure the system to provide continued service under these faults. Without this class of service, we could not provide the present-day high reliability LANs that are used for critical tasks such as real-time control, financial exchange, as well as many others.

The main function of these components is to keep the system functioning properly. It must do this in the face of both anticipated and unanticipated faults. Anticipated faults deal with those faults that the designers have thought of during the design of the system. These faults typically deal with hardware failures from which quantitative measures are available. Given failure modes (some limited number) and probability of failure, the most likely components to fail can be enumerated and the effect of their failure defined. This allows for an effective means of correcting or working around a failed component. Without such analysis of effect, appropriate workaround solutions could not be realized.

Anticipated faults and the solution to their effects have been well quantified. A large body of knowledge exists that provides us with many means to handle these conditions. As components get more complex, and the possible ways in which they can fail increase, we must learn to handle unanticipated faults—that is, conditions that were not envisioned from the start, and whose effects we do not or could not fully quantify, understand, or predict. Up to this point we looked solely at hardware failures and have developed fault-tolerant means to work around them. Hardware does not represent most faults in systems, though. Hardware can be well thought out and provided with built-in test points and hardware to test itself. Software, on the other hand, lacks the ability to be rigorously tested and as such represents a place susceptible to the unanticipated faults. We do not have ways to measure or estimate mean time between failure of software. Software fails due to erroneous design, due to faults in hardware, or due to other software that introduces data or conditions that will cause erroneous operation. To provide adequate protection against both hardware and software faults requires that we provide a means to test, monitor, and determine when errors occur. This chapter will cover the concept embodied in the monitoring, testing, detecting, and correction of errors in local area networks.

ERROR DETECTION

The starting point for any service that provides for error-free communications is found in detection. To correct errors or provide workarounds to keep processing requires that we can even determine that an error occurred. The better the error detection, the better and more reliable the end service will be. The tradeoff is in cost of added hardware and software to perform the monitoring.

Error detection techniques can be applied concurrently with processing, or as side checks to test operations off-line occasionally. The former is the more robust scheme, though it impacts performance (throughput); the latter is less costly, but only catches blatant faults—it will most likely not catch the intermittent errors. What kind of checks are adequate? How should they be used to correct, isolate errors? These are the questions that need to be answered in order to understand what performance monitoring constitutes and to determine what is a good method.

To provide good detection we must have checking of operations. Checks to be optimal must be based solely on the "black box" view of the device or software being tested. It should not be biased by internal idiosyncracies, which may slant a check in an erroneous direction. A check must be isolated from the entity being checked so as to provide isolation from a single-point failure (e.g., we do not want to use components from one entity as part of the check). The designer of the check should not be the same as the designer of the item to be checked again, to remove biases intended or otherwise.

Most LANs, though, may not have this luxury of being defined along with the test hardware and software. In reality, if these exist at all, they were added after providing less than optimal service. The checking done is typically based on providing some acceptable level. For a LAN, to detect that a message sent had an error, not where it was in the message, or to detect that a device on the LAN is flawed, but not actually isolate the flaw to any lower level, most performance monitoring checks are performed as output checks; that is, they provide testing of results for correctness based on known information.

The main forms of checking the correctness of results are based on:

- Replication
- Timing intervals
- Input/output reversals
- Encoding/decoding
- Completeness
- Structural
- Diagnostics

The first six forms of checking deal with testing the operational system for correctness during regular service; the last form uses testing off-line and infuses test data to check whether the proper results are seen.

Replication as a performance monitoring and detection scheme provides one of the most complete and powerful means to check for accuracy, although it also represents one of the most expensive ways to do the detection. Replication provides its service by having some form of copies of the entity being checked and a scheme by which the copies of outputs can be checked against each other for accuracy. The replication can be done for all components of a system or for only the critical ones. In a LAN, for example, we may wish to have two media used and two copies of network interface hardware and software.

The check would be to send the information over both links and compare them upon reception. If they match, the communication occurred correctly; if not, an error was detected. Other forms of replication do not include multiple hardware sets, but use multiple messages. If we wish to be sure a message is received without error, we could send multiple copies, have them compiled on the receiver and checked against each other; if they are all the same, no error; if not, use majority as correct message. The former replication provides a means to detect a wide array of errors whereas the latter deals with detecting transients.

Replication can be performed at whatever level is deemed necessary for proper operation. For a LAN, possibly the replication should be done on drivers, receivers, and media only—or possibly only on internal network interface-control hardware and software. The place to provide redundancy can be best chosen by performing a reliability computation to pick the pieces of the LAN that have the greatest potential problems, leaving the rest of the system alone. In this way one can maximize protection while minimizing costs.

Timing intervals provide a relatively inexpensive means to check the operations of a device or software. If the item of interest includes timing as a critical aspect of its operation, we could use this as a test point. If the timing constraint is missed, we signal that an error condition has occurred. Such checks are best used to indicate remote failures in LANs. For example, if a message is sent, and after some specified maximum time interval no response is received, then one can assume either the media, interface units, or the remote site has failed in some fashion. Using these potential conditions the sender can perform further tests to better isolate a problem, as will be seen in the next section of this chapter. Software timers can be used as checks of overall health. The monitored software is required on each of its operations loops to get a "watchdog" timer; if the timer is not set, it is assumed the software has failed. If it is set, all goes on.

A third form of detection is referred to as input/output reversals. This form of checking is based on reversing the processing of the output to provide the inputs back. So what occurs is that the outputs are put through a computation that provides as output the initial entities' inputs. These are then compared; if they match, a proper operation ensued, else an error occurred. The computation can be as simple as a substitution into intitial formulas, to as complex as actual recomputation of all processes back up to the inputs. These checks can only be performed where relationships exist between the input and output and they can be readily enumerated. This form of error detection is not used in LANs, at least not as a fine, granular tool. Typically, a LAN may utilize an echoing scheme, where the sender issues a message, the recipient will receive it and reissue it back to the sender. The sender can then check his original against the received message to see if the recipient got it correctly.

One of the most common forms of detection of errors in communications is the coding method. In this scheme the input is encoded with additional information that is derived from the original message. These code words or locations can be recomputed on the recipient's side. If they compute correctly, no error occurred, else one was detected. A common form of this end-to-end check is the parity check. Parity check uses the bits of the words being sent to compute parity bits.

The parity bits are computed as the modulo 2 sum of the inputted bits. The bit locations, along with a formula for interpreting whole words, are used to compute and then check the sent words. Beyond these simple checks are other more elaborate coding schemes such as binary block coding and cyclic redundancy codes. These codes and their deciphering can be computed and deciphered using simple hardware schemes. Due to the simplicity of the scheme and its hardware, these are typically used in LANs to check the integrity of transmissions. Details of coding can be found in Peterson [1972] and its references.

Completeness checking refers to a form of detection aimed at the isolation of test points and their collection and examination to determine if the outputs are reasonable and complete in relation to "expected" ranges of values. This type of checking provides a means to detect erroneous threads of execution, although it may not provide us with a means to determine who within the thread is faulty. Other tests may be needed to get this level of detection. This form of checking is similar to bounds-type checking performed by software. These provide the means to detect when faulty input data is being provided or faulty output data is being produced.

Structural checking is similar to the completeness checks in that it is applied to abstract data structures. The difference is that they are applied in a way to check that the intended structure is adhered to versus the bounds-type values. Typical checking would be to test the linkage of a data structure built from pointers to see if the first element indeed leads to the last. That is, in a ring the head element should fold back on itself when the ring has been fully traversed. This type of checking could provide us with a means in a LAN to test connectivity. We could send a route message that goes from node to node collecting status; on completion, the topology can be checked to see if it is as it is thought of. If not, then a node is down or in an error state.

The final form of checking is done as an off-line process. The diagnostic check is one where the system is stimulated with known data that should always, in a correctly operating system, provide a correct and a priori known output. Diagnostic checks provide a means to check hardware and software in such a way that error conditions or faults can be isolated. Diagnostics can be applied to test a LAN's control protocols, timings, queue buffers, memories, and other circuits. It can be used to test service programs and control programs. It provides the most complete way of exercising and testing for faults in the system.

Once we have test data to use we must have a means to use this data in determining faults and bringing them to the attention of proper management components. Performance-monitoring software provides services to collect information from various components of the system.

PERFORMANCE MONITORING

The performance monitor (PM) is also responsible for supporting the reconfiguration and recovery of the system through the detection and localization of system errors and hardware faults/failures to determine the proper reconfiguration/recovery procedure (reference Figure 4-1 for a high-level flow diagram).

Performance monitoring interfaces with and supports those functions responsible for informing the operator of the following needs:

- Warm/cold starts
- Software resets
- Subsystem reconfiguration
- Resource determination/reallocation
- Possible subsystem configurations

This component also provides subsystem alert reports to the systems availability manager indicating status on system errors, hardware faults/failures, or the loss of any functions/modes/submodes.

The PM mode is comprised of the following submodes:

- System PM executive
- System PM executive (secondary)
- Unit PM Executive
- Device services PM executive
- PM interface mapping executive

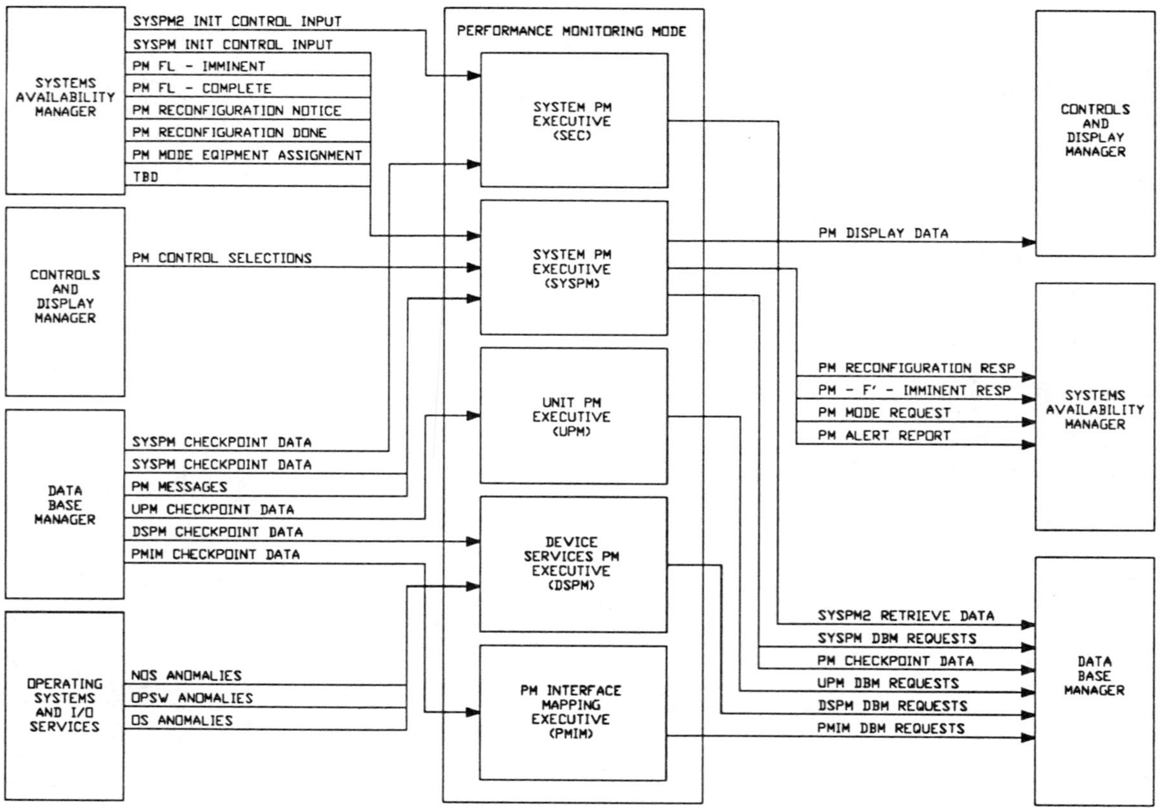

Figure 4-1. Performance-monitoring Mode High-level Diagram.

The system PM (SYSPM) Executive submode provides the control and interfaces necessary to allow for system-level performance-monitoring testing. SYSPM communicates with the systems availability manager to suspend PM testing prior to running fault localization (FL) and to resume PM testing upon completion of FL tests. It also interfaces with the Display Manager to provide PM display data to subsystem operators. The SYSPM stores, retrieves, and logs PM data via the Database Manager.

These interfaces provide SYSPM with the capability to schedule system-level testing, perform error correlation and network performance monitoring.

The SYSPM submode can be broken down to the areas discussed in the following paragraphs.

SYSPM Initialization is responsible for processing PM control requests and responses. This processing consists of accepting a PM suspend or resume request, forwarding that request to the respective subordinate process and updating the control event table. If the PM control request was to suspend or resume a subordinate process, this function will update the status report control table to accept or reject PM status reports to respective application processes. If the control request is for system-level PM tests, SYSPM will update the scheduling table to run or not to run the specified system-level PM tests. When all required processing has been completed, a response is sent to the requesting processor and all pertaining information is stored via the DBM.

The SYSPM Test Scheduling Function is responsible for the actual scheduling of all system-level PM testing. Additionally, it provides for the thresholding of system-level errors and building of status reports based on the results of system-level tests.

Thresholding of PM errors consists of reporting the specific PM test that initially detected the error to determine if the error is solid or intermittent. Results of this testing are compared against values stored within a threshold table containing predefined limits for each fault/failure. Once an error has been determined to be a solid fault or failure, it is added to the fault/failure history table and checkpointed to disk via the DBM. In addition, this data is sent to the SYSPM Error Correlation function for further processing.

SYSPM Error Correlation Function is responsible for analyzing related errors to determine the actual failing reconfigurable element. An example of error correlation would be the obtaining of power status to support data that inferred power supply problems detected by software checks or PM test results. Error Correlation is not necessary in all errors detected since the error may be isolated by the detecting PM test. This function, however, accepts error reports and PM status reports from subordinate processes and performs error correlation as required.

Once the error has been isolated to a reconfigurable element the following processes are performed:

- The PM error report is logged via the database manager.
- The system hardware status table is updated for reconfigurable elements to indicate fault or failure and checkpointed to disk via the database manager.

- The detailed PM message for the fault/failure is retrieved from the disk and associated with the failed reconfigurable element.
- The system availability manager (SAM) is informed via a PM alert report.

SYSPM Display Processing Function provides the necessary processing for the following PM displays:

1. System Hardware Status Displays. The SYSPM Deferred Repair Display is responsible for processing requests for and the generation of deferred repair information. Information concerning the data elements for deferral date and rationale is obtained from the operator and incorporated into the deferred repair table. Whenever an entry is added, deleted, or modified this information will be checkpointed to disk via the DBM.
2. SYSPM HW Status Table Display. This display function is responsible for generating and processing requests for hardware status table. The following operator selections would be available within this display:

   - Elements with faults or failures.
   - Elements for a specified unit with faults or failures.
   - Elements for a specified device type with faults or failures.

   Data elements within this table will include unit, device, type, date, status, fail time, modes, maintenance status, source, and associated PM messages.
3. Intermittent Error Display. This display function is responsible for generating and processing requests for the intermittent error display. The following operator selections are available within this display:

   - All elements.
   - All elements of a specified unit.
   - All elements for a specified device type.
   - Specified device.

   Data elements within this table will include unit, device, type, date, status, modes, maintenance status, time of failure, and associated PM messages.
4. Secondary SYSPM Executive. This submode provides all functions of the SYSPM Executive submode. It is initialized and brought online under the direction of the System Availability Manager in the case where the primary executive fails.
5. Unit PM Executive Submode. The Unit PM Executive submode provides interface between subordinate processes and the following:

   - SYSPM Executive—to report results of unit domain PM testing to receive and respond to PM control requests.
   - Device Services PM (DSPM) Executive—to control and obtain results of device PM testing.
   - Database Manager—for storing and retrieving PM test data.

The Unit PM (UPM) Executive is re-level error correlation of test data. The Unit PM Executive can be broken down to four functional areas:

- Initialization. Initialization is responsible for establishing the interfaces required via the network operating system (NOS). In addition, it retrieves UPM control data via the database manager.
- Test Scheduling. Test Scheduling is responsible for the scheduling of unit-level testing and for providing thresholding for errors detected. This information builds a status report based on the results of unit-level testing.
- PM Control. UPM PM Control is responsible for processing PM control requests and responses. As with the SYSPM control this processing consists of accepting PM suspend or resume requests and forwarding those requests to the subordinate processes. Upon completion of the PM test function, this function accepts control responses from the process and updates the control event table.

  If the control request is to resume or suspend unit-level PM tests, the scheduling table is updated to run or not run that specific unit-level test. A response is sent to the SYSPM when all required processing has been completed.
- Error Correlation. UPM Error Correlation is responsible for analyzing related errors and determining the failed reconfigurable element. Error Correlation operates under the same guidelines established in the section describing System PM Executive Error Correlation. Once UPM status reports are generated they are forwarded to SYSPM for further processing.

6. Device Services PM Executive (DSPM) Submode. This submode is responsible for providing interface between:

- Unit Executives (UPM)—provides status of device-level PM testing, responds to PM control requests.
- Network Operating System (NOS)—gathers NOS-detected anomalies.
- Systems Operation System—gathers SOS-detected anomalies.
- Operating Software—gathers operational software anomalies.
- Database Manager—stores and retrieves PM data.

In addition, the DSPM is responsible for scheduling device-level PM testing and providing device-level error-correlation of test data. The DSPM can be broken down to five functional areas:

- Initialization. DSPM Initialization is responsible for establishing the interfaces required via the NOS. In addition it retrieves DSPM control data via the DBM.
- Test Scheduling. DSPM Test Scheduling is responsible for scheduling device-level PM testing and providing thresholding of errors detected. With this information it builds a status report providing the results of Device-level testing.

- Error Correlation. DSPM Error Correlation is responsible for analyzing related errors and determining the failing reconfigurable element. This error-correlation process operates under the same guidelines established in the section covering SYSPM Executive submode error correlation. Once a DSPM status report is generated it is forwarded to the SYSPM for further processing.
- DSPM Control. DSPM PM Control is responsible for processing PM Control requests and responses. As with SYSPM control, this processing consists of accepting a PM suspend or resume request and forwarding that request to the subordinate processes. If the Control request is to resume or suspend device-level PM testing, the scheduling table is updated to run or not run the specified PM testing. A response is sent to the respective UPM when all required processing is completed.
- DSPM Tests.

7. PM Interface Mapping (PMIM) Executive Submode. This submode is responsible for providing interfaces with the following:

   - SYSPM—to report the results of the retained and certain contractor-supplied unit PM testing and to receive/respond to PM Control requests.
   - Database Manager—for the storage and retrieval of test data.

   PMIM is responsible for the scheduling of interface handlers to initiate (if required) and retrieve the results of retained-unit PM tests and provide error correlation. PMIM can be broken down into four functional areas:

   - Initialization. Initialization is responsible for establishing network interfaces required via the NOS. In addition, it provides necessary interfaces with the DBM to store and retrieve data.
   - Test Scheduling. Test Scheduling is responsible for scheduling interface handlers to initiate (if required) and retrieve results of the retained and certain contractor-supplied unit PM tests.
   - Error Correlation. PMIM Error Correlation is responsible for analyzing related errors and determining the failing reconfigurable element. This function operates under the same guidelines established in the section covering System PM Executive Error Correlation.
   - PM Control. PMIM Control is responsible for processing PM Control requests to suspend or resume testing. Not all retained-unit PMs support these control requests and in those cases the response will indicate this. If the control request is to resume or suspend retained-unit testing, the scheduling table is updated to run or not run that specific interface handler for those retained-unit tests. A response is sent to SYSPM when all processing has been completed.

The performance monitoring functions described represent a superset, and as such one would not see all these in any one system unless it was a high-reliability fault tolerant design.

Fault Isolation

Performance monitoring provides a base of information to help define what has failed in a system. The goal of performance monitoring is to detect errors and collect data on them. Fault isolation utilizes this data along with further tests to isolate the causes of PM-detected faults. In other words, the major function of the fault isolation component is to diagnose the cause of a problem down to an acceptable level of granularity. In some systems this granularity may be a processor, or channel, an I/O and/or any printed-circuit board in the system, as well as any component on the circuit board. The issue is level of granularity of detection and isolation and how long you are willing to bring your system, or individual computer, off line to define and isolate a hard fault in the hardware or software.

Typical fault-isolation packages operate in a hierarchical form, first running simple localization tests using the known performance-monitoring data to isolate the error conditions down to some major device or subsystem. Once this localization of an error has been performed, the fault-isolation services will begin iterative diagnostic tests to further isolate a localized fault. The process for doing this typically consists of applying known input data to a software or hardware component, and stepping through the procedure, examining test data at every possible point. This data is then correlated with known expected outputs on the steps to localize the error condition. Once the error is localized, fault-assessment algorithms are performed to further define the impact of this error on other components. In this fashion the system faults can be found and recommendations/procedures to correct them performed. Within a LAN environment, fault isolation deals mainly with detecting and localizing a flawed network component, then determining the part or parts responsible for the error, as well as assessing how this fault will impact overall system performance. Fault isolation in most systems utilizes additional equipment that can be attached to the system. As an example of a fault isolation hierarchy the following system (Fortier 86) is defined.

Fault Localization

Fault Localization is the primary maintenance tool for organization-level maintenance within DSDB. It will provide automated testing of off-line hardware devices utilizing Common Action Entry Panels (CAEP) for the man/machine interface, ROM-based or disk-stored FL test programs and FL control programs.

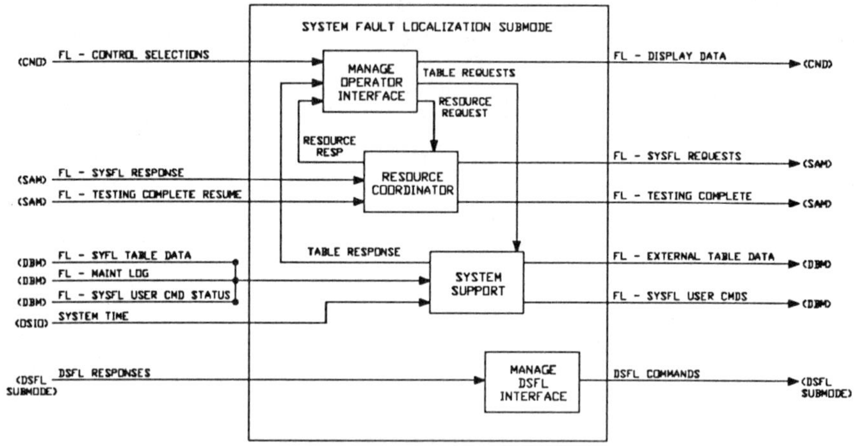

Figure 4-2. System Fault Localization Submode.

To complete this task, the Fault Localization mode uses four submodes:

- System Fault Localization (SYSFL)
- Device Services Fault Localization (DSFL)
- Test Processor
- Test Programs

System Fault Localization Submode (SYSFL) SYSFL is the highest level of FL control and has the capability of functioning from any major processing element. It verifies FL test selections, verifies that the required hardware is the proper on-line or off-line condition, passes operator requests to the Device Services FL executives (DSFL), and accepts responses from DSFL and displays them to the operator (reference Figure 4-2).

SYSFL consists of four processes:

1. Manage Operator Interface that manages operator FL test selection, initialization, and control from control panels simultaneously.
2. Resource Coordination notifies the System Availability Manager mode of needed resources and categorizes the devices into Sensor Devices and Real-Time Control Devices.
3. SYSFL System Support provides the interface to the Database Manager mode. Commands to the DBM are formulated based on function requests.
4. Manage Device Services Fault Localization (DSFL) Submode Interface provides the necessary control logic to manage one active DSFL per panel at a time.

Device Services Fault Localization DSFL and associated test processors are used to control a set of FL tests against a specific device. DSFL receives commands from SYSFL, loads and activates the test processor, and places command codes in an area known as interface storage. The Test Processor then reads the commands sent from DSFL and acts on them (Figure 4-3).

DSFL consists of the following processing:

1. Manage SYSFL Interface provides the interface between SYSFL and DSFL used for the passing of commands and responses.
2. DSFL System Support provides interfaces to other System Service Modes.
3. DSFL Create Test Process creates and aborts test processors on command from SYSFL. On notification from Manage SYSFL Interface that a SELECT FL has been received, the loading address of the test processor is determined and a test processor load request is sent to the DSFL system support component.
4. DSFL Manage IF Storage communicates to the Test Processor submode using two mechanisms. The primary mechanism is passive and uses the interface storage area. A secondary mechanism used is the Executive Service Requests of the host operating system.

Test Processor Submode The Test Processor submode communicates with and is controlled by the DSFL submode. DSFL and Test Processor together form the low-level FL test executive from which FL tests are controlled. The many different test processors throughout the system allow both new and retained equipment to be controlled by DSFLs and by SYSFL (Figure 4-4).

Test Processor consists of two processing functions:

1. TP Manage IF Storage receives and validates commands from DSFL and returns validated commands to Test Control.
2. Test Control provides the level of test program control appropriate to the device under test.

Test Programs Submode Test Programs submode is the collection of FL tests that have been written to test a specific hardware device. The programs provide the following general functions (Figure 4-5):

- Initialize hardware under test.
- Stimulate the hardware under test.
- Collect responses from the hardware under test.
- Determine pass/fail status.
- Provide test results.
- Provide for operator support if needed.

As was indicated, the fault-isolation component of a LAN or any system transport and monitoring subsystem has the job of off-line determining what errors occurred,

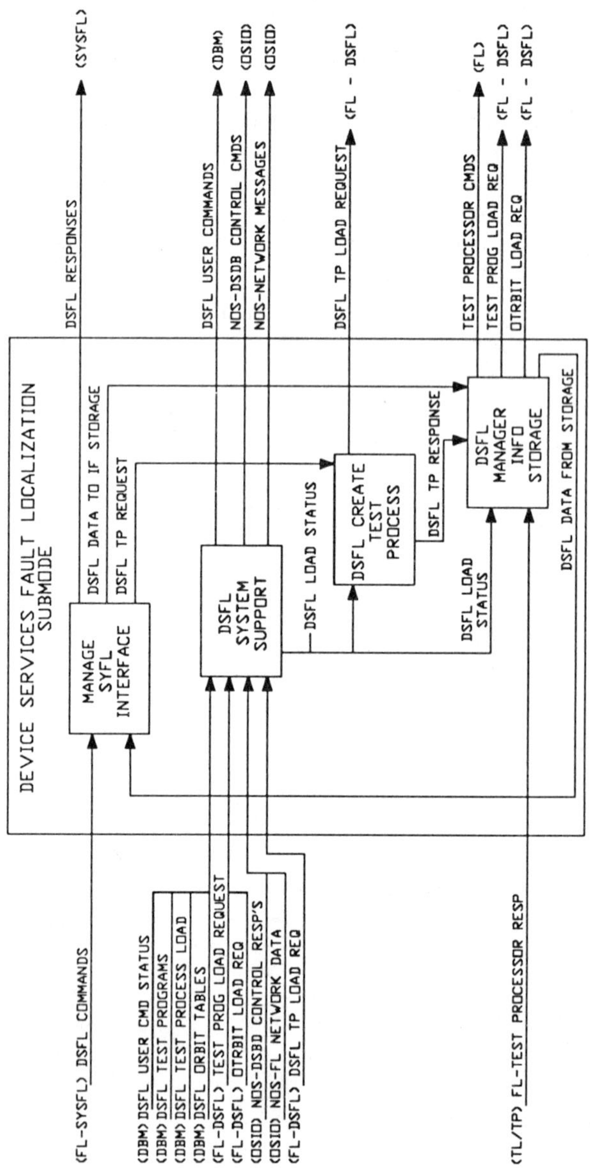

Figure 4-3. Device Services Fault Localization Submode.

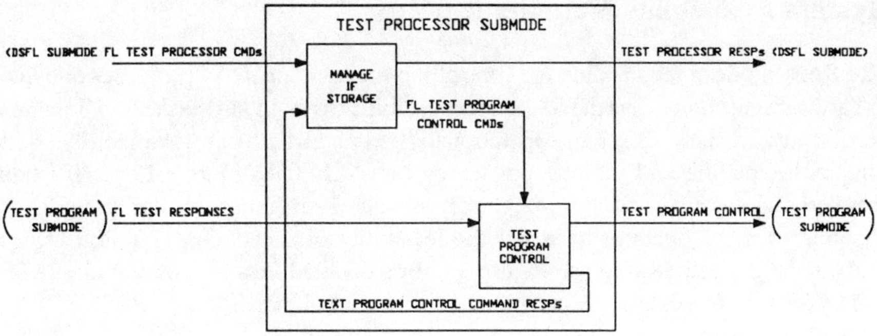

Figure 4-4. Test Processor Submode.

where they are located, and to assess their meaning and impact with the rest of the system.

SYSTEMS MANAGEMENT

Beyond the low-level error detection and isolation functions lies systems management. In this context systems management does not refer to operating systems or data management, but to systems assessment of configuration and status of the entire LAN hardware and software suite. This service component of a LAN consists of systems availability management that determines and manages systems loads, configuration management that maintains the status of the present loaded system (hardware and software), network communications management that maintains and manages communications assets, and reconfiguration management that provides the means to recover the system to a working state after failure.

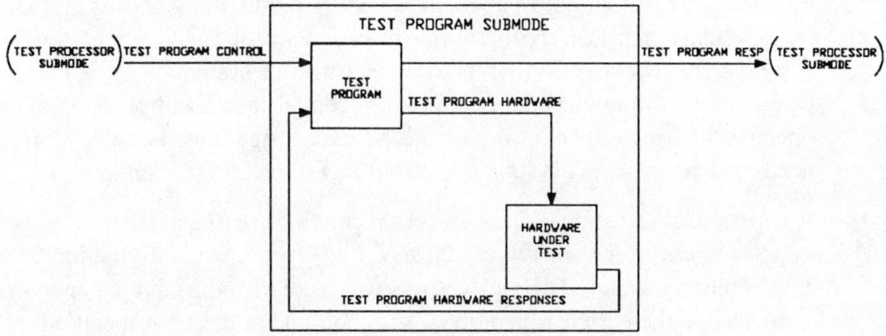

Figure 4-5. Test Program Submode.

System Availability Manager (SAM)

The SAM mode is responsible for controlling the assignment of processes and tasks to hardware resources. It is the SAM's responsibility to maintain the desired functional capabilities to meet changing operational needs and hardware availability. SAM directs the loading and initiation of the system at Initial Program Load (IPL) and recovers the system following partial or entire system failure. It also provides the capability for the operator to assess the feasibility of a new configuration prior to making the change. (Reference Figure 4-6 for a detailed flow diagram of the SAM.)

The System Availability Manager consists of six submodes:

- System Initialization
- Operation Configuration Set Selection
- Tactical Model Selection
- Error Recovery
- Offline/Online Status Handling
- Reconfiguration Execution

System Initialization Submode The System Initialization Submode loads and initializes the system following a Power On Restart (causing an IPL) and following a request for a system warmstart or coldstart.

The four components of this submode are:

1. Local Area Network Initial Program Load (LANIPL), which loads the NOS functions into individual LAN elements. On completion, the loaded functions support the loading, initialization, and process-to-process communication application software. System Services Initial Program Load (SSIPL) loads and initializes the systems availability manager and the configuration manager. System Startup loads all devices in the system with application programs and starts these programs.
2. System Takedown causes the loadable devices in the system to reach a Power-On-Restart state to prepare them for loading. System Takedown is initiated by a system coldstart or warmstart request from Controls and Displays Management. On receipt of this request, System Takedown performs a systematic takedown using resets to the BIUs of the system.
3. Operation Configuration Set Selection Process accesses and presents to the operator the feasibility of changing the system configuration to reflect current needs and priorities. This submode consists of three processes:

    - Operation Assessment. This process receives a command from the supervisor to assess the feasibility of a specified Operation Configuration Set.
    - Operation Change. This process receives a command from the supervisor to change the current Operation Configuration Set to a new specified set.
    - Operation Definition. This process interfaces with the supervisor and allows him to build three custom-option configuration sets.

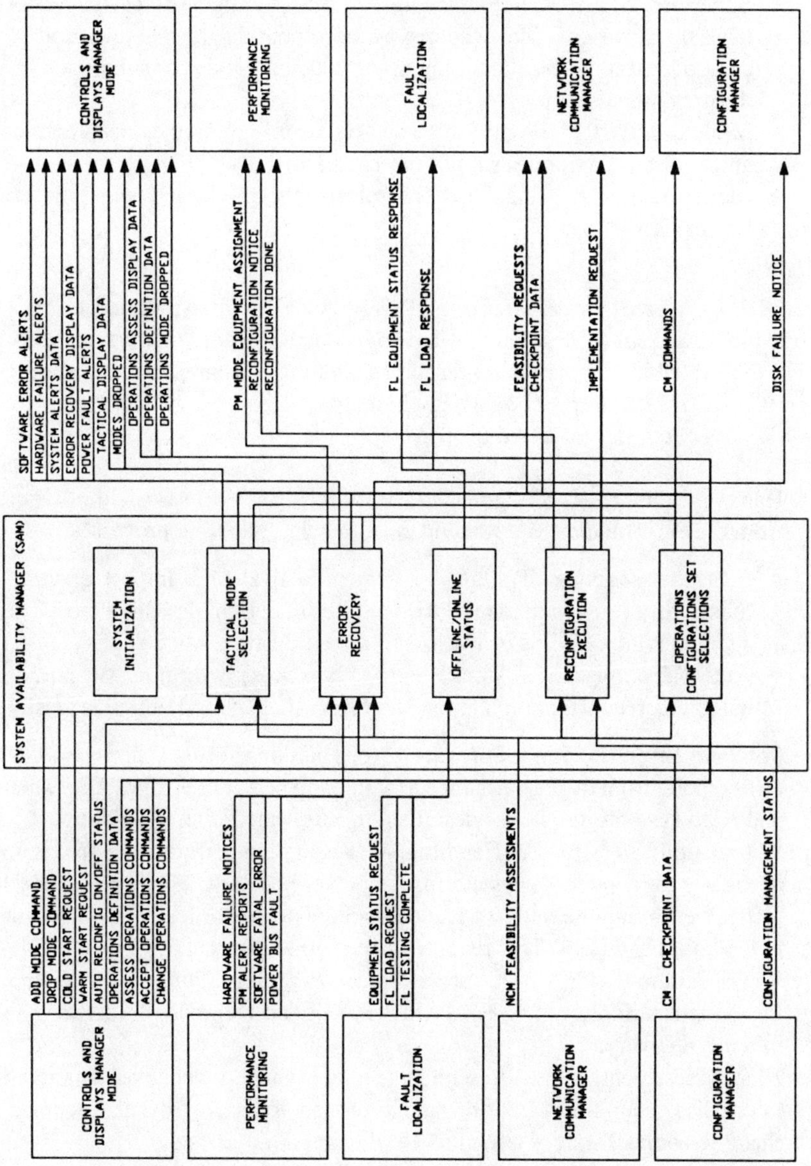

Figure 4-6. SAM Interfaces and Functions.

4. Applications Mode Selection allows the supervisor to add, drop, warmstart, or coldstart modes in the system. Three processing functions comprise the Applications Mode Selection. They are:

   - Supervisor Mode Change. The Supervisor Mode Change receives requests from the supervisor via Resource Management displays to add modes or submodes to the operational lineup or to delete modes or submodes from the operational lineup.
   - Mode Coldstart. Mode coldstart executes reconfiguration requests to reload and coldstart the modes supporting processors.
   - Mode Warmstart. This functions reloads the mode and warmstarts its processes.

Error Recovery assesses the impact of Performance Monitor-reported errors on operating modes and submodes. It initiates recovery actions based on the type of error reported, the availability of hardware resources, and the current submode priorities. An alert is sent to the Controls and Display Manager.

Error Recovery consists of seven processes:

1. Hardware Failure Handling. This process receives hardware failure reports from the Performance Monitor and performs the following processes:

   - Failure Assessment. Failure Assessment analyzes the impact of Performance Monitor-reported hardware failures to determine which modes and submodes are affected by the loss of the hardware resource.
   - Auto Reconfiguration Assessment. This process determines whether an automatic reconfiguration to recovery from the reported failure is feasible.

2. Software Error Handling. Software Error Handling receives software fatal-error reports from the Performance Monitor and performs Failure Assessment and Auto Reconfiguration as described in Hardware Failure Handling.
3. Automatic Power Transfer Handling. APT Handling performs the processing necessary to recover the system from a power bus fault. The processors that lost power during the fault are loaded with the processor loads they had prior to the fault conditions. This recovery is performed automatically.
4. Power Failure Handling. This process receives power failure reports from the Performance Monitor. It assesses which modes and submodes are affected and initiates recovery.
5. Alert Assessment. Alert Assessment generates an alert whenever a hardware failure has occurred. In addition, an alert is sent to the supervisor whenever a Spare Resource Limit or a Critical Hardware Failure Exists.
6. Self-Recovery. Self-recovery ensures that the operation of SAM is interrupted for no more than two seconds in the event of power bus failure, or a failure of the processor or processor load in which the SAM program resides. Self-re-

covery retains a history of the system configuration and hardware status immediately preceding the interruption.

7. Error Recovery Option Computation. This process computes which modes will remain in the system (ON) and which modes will be removed from the system (OFF) based on an equipment failure. Reconfiguration options are presented to the supervisor for interaction and approval.

Off-line/On-line Status Handling processes requests from the Resources Manager Displays to take a piece of equipment off-line. SAM calculates the impact of taking the equipment off-line and recommends which modes to drop. The operator will be given the option to proceed with the off-line request or cancel it.

Reconfiguration Execution communicates with the Configuration Manager mode to provide the latter with the load, delete, start, and stop commands necessary to carry out the planned reconfiguration.

Configuration Manager Mode

The Configuration Manager is responsible for loading the various application programs into their host processors under the control of the SAM. The CM mode also maintains communications with both application and executive programs to manage the starting, stopping, and deleting of various processes within those programs. Figure 4-7 illustrates a high-level diagram of the CM mode.

The CM is comprised of three submodes:

- Process Load
- Process Control
- Checkpoint

Process Load Submode The Process Load submode of the CM mode is responsible for loading the various host processors with the processes designated by the SAM. On completion it returns status information concerning the successful or unsuccessful execution of the load command to SAM.

The Process Load submode can be broken down into the following four functions (reference Figure 4-8):

1. Initialize Processor Function. The Initialize Processor function is responsible for initializing processors as directed by the SAM. This initialization consists of the preparation of a processor to be loaded and the return of completion status to SAM after executing the command.
2. Load Process Function. The Load Process function is responsible for loading of application software into initialized processors under the direction of the

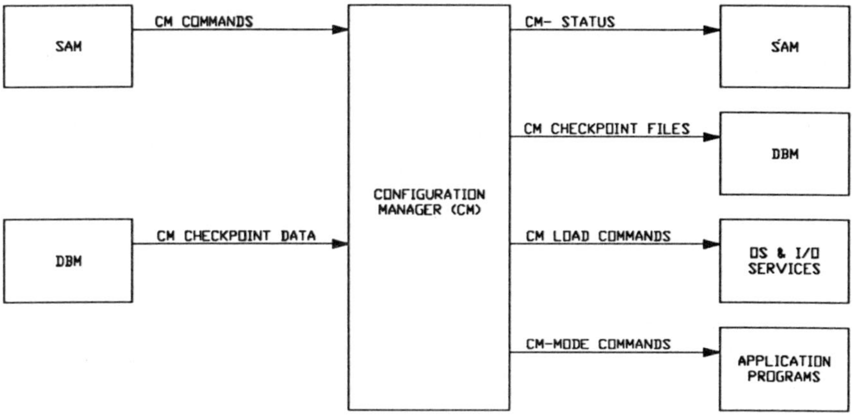

Figure 4-7. Configuration Manager High-Level Flow Diagram.

SAM. On completion of the respective load, the load status (successful/unsuccessful) is returned to the SAM.

3. Delete Process Function. The Delete Process function is responsible for deletion of individual processes in accordance with instructions issued by the SAM. After execution of the Delete Process command, the deleted process is not replaced by itself or any other process until a complete reload of that processor is instructed by the SAM.
4. Verify Logic Function. The Verify Logic function verifies that a specific Bus Interface Unit (BIU) is loaded and returns this status to SAM on request.

Process Control Submode The Process Control submode is responsible for controlling the starting and stopping of software processes. In addition, this function provides the capability for starting, stopping, or resetting any or all of the subsets of a process.

The Process Control submode can be broken down into three functions (see Figure 4-8):

1. Mode Start Logic Function. The Mode Start Logic function controls the starting of a mode or submode under direction of the SAM. This is accomplished by issuing a start command to the controlling processor of the respective mode or submode to be started. Two types of mode start commands can be issued:

   - Warm Start Logic—provides control to the respective application processor to start its functions utilizing the latest checkpoint parameters.
   - Cold Start Logic—provides control to the respective application processor to start its functions utilizing specific default operational parameters.

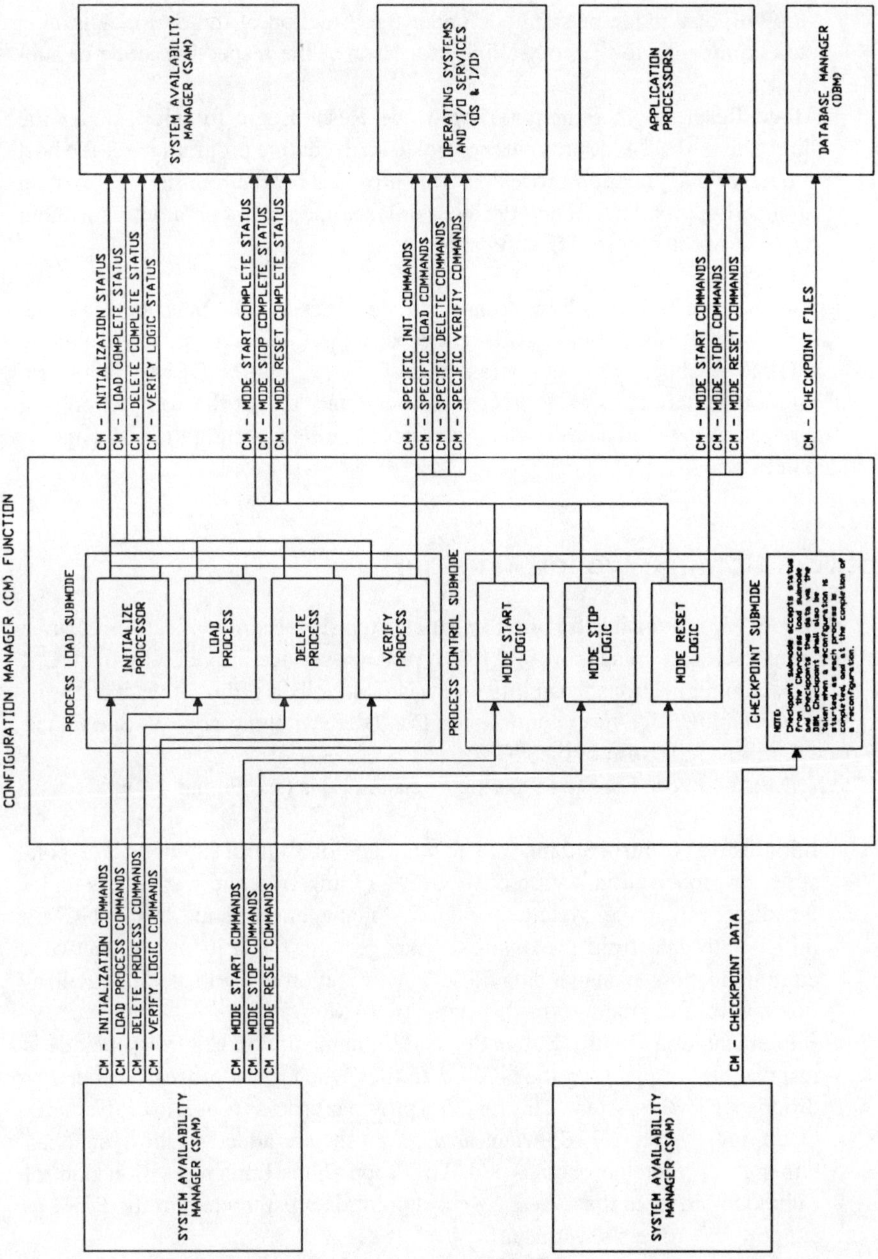

Figure 4-8. Detailed Flow Diagram of Configuration Manager.

2. Mode Stop Logic Function. The Mode Stop Logic function controls the stopping of a mode or submode under the direction of the SAM by issuing stop commands to the controlling processor of the respective mode or submode.
3. Mode Reset Logic Function. The Mode Reset Logic function, under the direction of the SAM, provides control to an executive program to reset a host processor's application processes. This provides for the reinitialization of an application process utilizing the last available checkpoint parameters without a reload of the application software.

Checkpoint Submode The Checkpoint submode is responsible for accepting status from the Process Load Submode and checkpointing this data via the Database Manager (DBM). Checkpoint data parameters are stored via the DBM whenever a reconfiguration is started, as each process is completed, and at the completion of a reconfiguration. A detailed data flow diagram representing Configuration Manager is provided in Figure 4-8.

Network Communication Manager

The NCM mode is responsible for providing the system-level communication resource management necessary to support SAM. The primary service provided by the NCM are network communications feasibility analysis, Bridge/Bus failure impact analysis, Bus Interface Unit/Bridge reset control, and Bus BIU switching control. See Figure 4-9 for a high-level diagram of the NCM.

The NCM is composed of the following five submodes (see Figure 4-9):

1. Initialization Control Submode. The Initialization Control submode is responsible for proper initialization of the NCM during cold and warmstarts under the direction of the system availability manager. It requests the NCM's initial-state data from the database manager and loads it into the network communications manager data files. On completion of the load, the initialization control function returns this status to SAM.
2. Implementation Control Submode. The Implementation Control submode is responsible for updating the NCM data files whenever a process is added or dropped from the system. This function provides process routes to the Network Operating System (NOS) when new processes are added to the system. In either case, when a process is added or dropped, the Implementation Control Submode provides the necessary checkpoint data parameters to the DBM to support the process's warmstart.

   This function of the NCM is responsible for receiving routing responses from the NOS, indicating the success of putting each routing parameter in place, and returning the status to SAM.

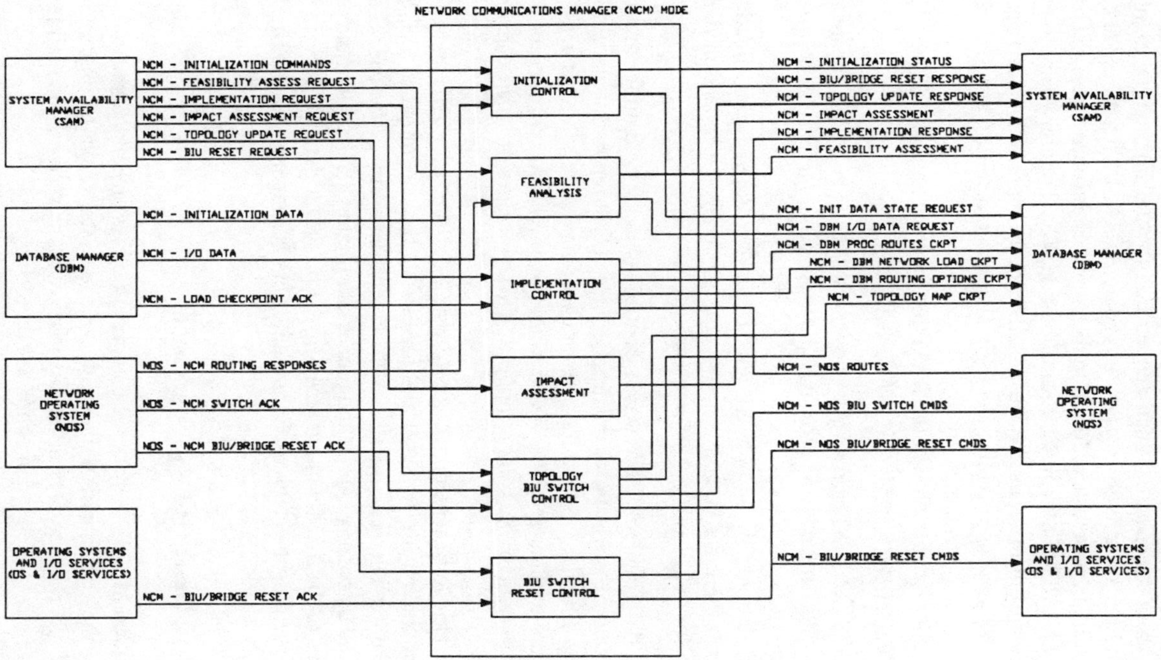

Figure 4-9. Flow Diagram of Network Communications Manager.

3. Impact Assessment Control Submode. The Impact Assessment Control submode is responsible for determining the impact of routing an application process through a specific bridge or bus domain. This is accomplished by checking whether specified bridges or buses are listed as active within the NCM topology map. When active, the program ID that is currently routing message data is returned as status updates to SAM. This submode can support the Impact Assessment for up to four network resources per request.
4. Topology/BIU Switch Control Submode. The Topology/Bus Interface Unit Switch Control submode is responsible for the management of Topology Map updates and for providing the logic necessary to control the Array Bus BIU's active legs.

   Under the direction of the SAM this function adds or drops bridge and bus domains from the active topology list:

   - Additions: When network bridge/buses are to be added, the information is placed in the NCM data files and tagged as available for communications services.
   - Deletions: When network allocations are dropped, this function ensures that there are no active application processes dependent on that resource prior to executing the command.

   If the Topology/BIU Switch Control submode determines that there is an active process related to the drop request, it will return this status to the SAM along with the program ID without changing the Topology state.

   For any change to the NCM topology, the NCM will modify the routing topology map and routing options. These changes are then checkpointed to the DBM.
5. BIU/Bridge Reset Control is responsible for accepting BIU and Bridge reset commands from the SAM and forwarding those commands to the NOS for execution. On receiving Bridge Reset commands, the NCM checks the Bridge Active list for any application processes utilizing that resource and returns that status to the SAM without resetting the bridge. A detailed data flow diagram representing the Network Communications Manager is provided in Figure 4-9.

RECONFIGURATION MANAGER

Reconfiguration Management is the component of systems management responsible for utilizing the configuration management and systems availability management, PM/FL, and other data to determine how to recover the system, upon failure, to some consistent state by which it can continue processing.

To perform this function, the reconfiguration manager must determine what processes, tasks, and hardware are affected, and which pieces of these affected components must be brought back on line. If the system can handle it, the desirable thing would be to bring all affected processes and tasks back up. If not, the reconfiguration manager must decide which tasks or processes to shed in order to keep the most important ones functioning. Reconfiguration occurs for three main reasons: normal system rebuilds and restarts, casualty reconfiguration, and situation-based events.

Reconfiguration is more of a philosophy than an actual physical entity. It pervades all components and has pieces everywhere. When taken together, they will provide the reconfiguration. In the architecture disscussed in this chapter we assume reconfiguration is actually embedded in the systems availability manager. Before we discuss an architecture, the definition of reconfiguration classes would help.

Normal reconfiguration, or nonerror reconfiguration, occurs when the user (systems operations) selects the configuration of the system to be brought up. The reconfiguration system has the job on this condition to check the present status of the system and determine if this reconfiguration is possible. If so, it uses other systems' services to effect the reconfiguration. If not, an operator must be brought into the loop to determine what must be shed to allow this reconfiguration to occur.

The second form of reconfiguration deals with fault conditions. When faults get to a threshold level, where system-critical activities are being impaired, or user applications are forced to wait, abort, or be indefinitely impaired, casualty reconfiguration enters. This form of reconfiguration goes through a process (Figure 4-10) that requires the assessment of the casualty, the construction of a solution (reconfiguration), the validation of the solution, and the implementation. Casualty reconfiguration occurs when a processing resource fails or a software component fails. In either case, the procedure is the same, though the degree of work that must be performed is different. Recovery from a hard fault (hardware) is called reconfiguration, whereas recovery from a software fault is simply software recovery.

The last form of reconfiguration is referred to as a Rapid Situation Reconfiguration. This form of reconfiguration would be found in critical real-time control systems where, based on event conditions, the system automatically reconfigures to another state to handle the situation.

The typical scenario for reconfiguration shown in Figure 4-10 follows:

1. The performance-monitoring subsystem detects errors and informs a systems correlation function as to their existence and degree of failure.
2. The system correlation process converses with the systems availability manager to determine damage and to prepare the damaged nodes for a restart.
3. The systems availability manager takes the information on errors and determines the extent of the impact on the systems operations.
4. If the assessment is such that it warrants a reconfiguration, then the systems availability manager determines a set of feasible solutions.
5. These solutions are checked versus the present loads and network capabilities to determine which of the solutions is the most optimal to institute.
6. The selected solution is validated by a check process and if necessary via operator approval.
7. The approval reconfiguration is then implemented using the facilities of the configuration manager.

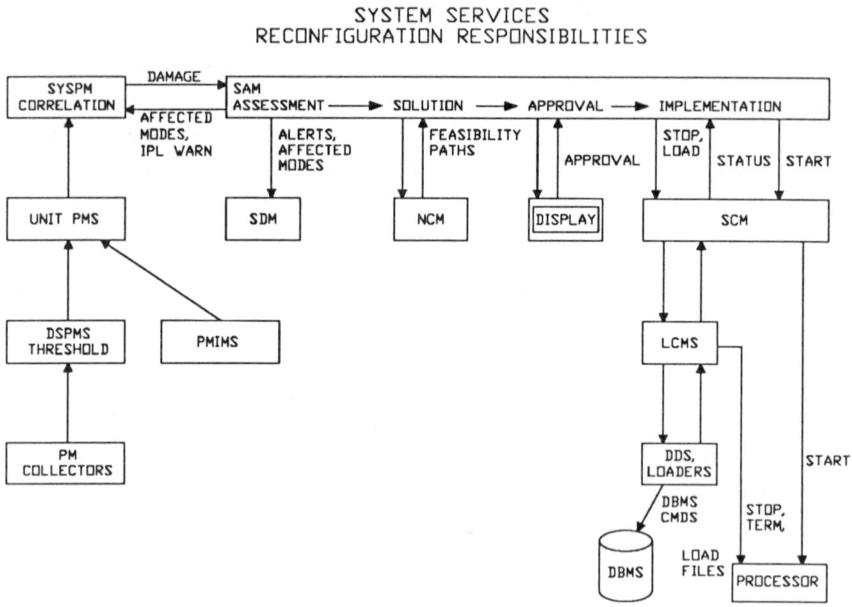

Figure 4-10. System Services Reconfiguration Responsibilities.

All reconfigurations, whether casualty or otherwise, go through similar assessments to determine the impact of reconfiguration.

Another aspect of reconfiguration is software recovery. In the event of a software fault, we have three possibilities:

1. The fault destroyed the program.
2. The fault destroyed data.
3. The fault destroyed both.

Based on the error conditions three forms of software recovery can be performed: coldstart, warmstart, reset.

On reset we assume that the data was erroneous, but the software is OK. To reset the software we reinitialize any data values to a state that existed before the error, then we download new data. The program is restarted and processing continues.

In a warmstart situation the program and its data fail, but the historical data being processed is okay. In this case, we need to download both programs and state data, then restore the program to a known state before the error (checkpoint data is used for this), then restart the program from this point.

Coldstart is performed when total failure has occurred. Historical data as well as program and data state information is lost. In this case, we must download all data files, programs, and state data and restart the entire program from scratch. This process requires the use of the other management modes to effect the operation.

SUMMARY

This chapter introduced the concepts embodied in the management and control of errors in the LAN environment as well as in any computing environment. The concepts discussed are for "full-powered" systems and, as such, not all these features are required nor are necessary in many LAN environments. The decision a LAN developer or purchaser must make in selecting which if any of these features to include is based on the degree of reliability one wishes to have within the network, and the degree to which one wishes to offload the detection and correction or handling of errors to the user. Readers interested in further details of systems that utilize facilities such as this are directed to Anderson and Lee's text on Fault Tolerance.

5

DATA LINK/NETWORK

INTRODUCTION

The data link/network layer provides the low-level services for routing of messages over the LAN, for controlling the flow of data over the LAN. It provides for addressing capabilities and for synchronization and error-handling capabilities, as well as the low-level message handing. Additionally, the function of this layer is to provide services to access other LANs via bridges and gateways.

ROUTING/FLOW CONTROL

Routing is the mechanism by which messages are transported through a network. The route is the "path" through the network that a message traverses. Routing algorithms operate by choosing the proper path to send a message that minimizes or maximizes some heuristic. The path selected can be fixed physically and held for the duration of a process to process session, or it can be variable and held only for a packet or message at a time. The routing scheme used is dependent somewhat on the forms of data transfer service offered. For example, if we offer a virtual circuit form of linkage, then the routing scheme must provide for linkage held over a time period, whereas if we have a datagram-type service, the router can choose any means to get the packet out, since no guarantees on the message are implied.

The virtual circuits represent dedicated paths set up between the entities involved in a session. These "circuits" exist for the duration of the interaction; therefore, they require that the routing algorithm be responsive to the requirements of delivering packets/messages in order, and that it interact/utilize other network services to guarantee "correct" error-free transmissions.

In the latter case, the datagram-type service makes no attempt and provides no service to guarantee correct and error-free transmissions. The router then is left to its own scruples to determine how to attempt transmission. It does not need to interact with any additional services; it simply determines a path, then sends the message out based on its protocol.

The difference in LAN environments versus wide area or metropolitan area networks in terms of routing is simplicity. LAN routing schemes typically are built into the design by media topology choice. That is, the LAN topology drives the need for routing algorithms. For example, the bus topology of an Ethernet-type LAN has a routing scheme called "broadcast" or "flooding" in ARPAnet terms. The sender simply broadcasts the message out with an address attached to it. As in flooding, all units hear it; the one with the matching address takes it off the LAN for use. The others listen, but do not copy it. Another example of topology determining routing is the simple ring. The ring topology by design forces the sender to always send out (route) a message using the single-output link. There is no other means to send messages out utilizing this topology. The need for true routing algorithms arises in those topologies where multiple media are used (e.g., dual ring, dual bus), or when multiple networks are connected (via bridges or gateways). In those cases, the LAN must utilize a scheme to route the messages over to the appropriate LAN and ultimately the destination node.

The LAN star topology offers a situation where routing is needed. The central hub receives all traffic and must determine how to send it out. The protocol to select a link could be a simple flooding algorithm, or an intelligent one where the hub examines the address and determines which link to send it over.

A LAN more closely related to the predecessor wide area networks is the mesh or irregular network. They have topologies that match the complexity of the WANs and require much the same forms of routing and control.

Routing schemes operate in two phases; the first phase is an examination of the destination address to determine the direction. The second phase is to select a path to send out the message that heads it in the right direction.

The major types of routing algorithms found in the literature include:

- Flooding
- Static table routing
- Adaptive routing
- Centralized
- Isolated
- Distributed

Flooding

In this algorithm for routing, each node that receives a packet/message first determines whether it has seen this before, and if it has, it discards it; if not, it forwards it out to

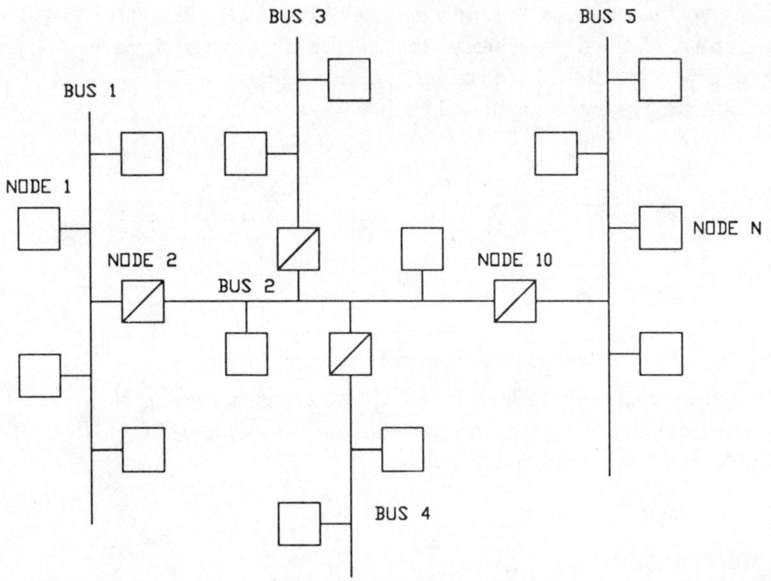

Figure 5-1. Multibus Topology.

all the neighbors except the one(s) it arrived from. In the case of Figure 5-1, if node 1 wishes to send a message to node N, it would ship the message out ("broadcast") onto the global bus 1 that it is connected to. In turn, this would be received by all the nodes on this bus who would then, based on how they are connected, forward this on. In this case, node 2 would forward this message on to its connected subnetworks. Ultimately, the message would get to the destination node N by traversing to node 10, who in turn would broadcast the message over bus 5 to node N.

This form of routing is extremely robust; that is, it will always work under all conditions. Additionally, it is guaranteed to find the optimal path from any source to any destination. The problem is that due to this robustness, we have added message loading that otherwise is unneeded. For example, again using Figure 5-1, if we assume the same scenario as above, we are forcing the three uninvolved subnets also to be burdened with the extra message from node 1 to node N, even though they do not need it. This is all right if we have low utilizations on the networks; but as traffic increases, all subnets will feel the impact.

Static Routing

A second form of routing alluded to earlier is static routing, also known as directory routing. This form of routing requires that a predefined routing table be specified and exists for each source/destination pair in the network. Additionally, this static table does not adjust to changing conditions within the network.

In the previous example, to send a message from node 1 to node N would require looking up a predefined route through the networks. In this case, broadcast on network 1 to node 3 which will then send a message to node 10 to node N.

The table for this entry may look like this:

	DESTINATION NODE					
SOURCE NODE	1	,	,	, , ,	,	N
1						3
3						10
10						N

This shows what node to send to in order to ultimately get to N. Node 3 in turn would have an entry of source 3 and destination 10 to send to Bus 5. Node 10 would have an N, indicating a direct path exists.

Adaptive Routing

The third major class of routing algorithms is referred to as adaptive strategies. These strategies can change and adjust according to the traffic flow and availability conditions in the LAN. These types of schemes would be beneficial in applications where reliability is put ahead of cost and timing. That is, to perform these dynamic schemes will cost more in terms of design/maintenance, etc.

There are basically three basic types:

- Centralized
- Isolated
- Distributed

Centralized adaptive routing utilizes a central routing control center that takes in systems status information (failures, traffic, etc.) and computes new routes for each node. These new routes are downloaded to the sites and used as their new routing tables. The issue becomes one of how often to do this and by what parameters are the new routes decided on.

Isolated routing is a scheme by which each node utilizes its own knowledge to base routing decisions on. An example of an algorithm of this class is the hot-potato algorithm. This algorithm operates by taking incoming messages, and outputting them as fast as it can on another link. Preferably, it would be the lowest-utilized link or a link with the shortest output queue.

Distributed routing is a form of routing where each node exchanges information with its neighbors (in the case of a global bus LAN this may be all the nodes on the bus). This information is then used to calculate optimal routes.

In the LAN environment this may mean that each node tells all others who it is connected to. This in turn can be sent to the other nodes so that they can build a view of the topology and a routing table from this information.

FLOW AND CONGESTION CONTROL

Flow control and congestion control are needed in networks where excessive message loading causes loss of messages and degraded performance.

Congestion is the fault and flow control is a preventive means to halt congestion. Congestion is the problem of more messages arriving at a node than it can hold. This is a node-to-node phenomenon.

The problem with congestion in a LAN or other network is that it will degrade performance (transmission delays will increase), reduce throughput, and can lead to deadlock in the system.

Congestion is caused mainly by four conditions:

1. Network interface units are too slow to perform their tasks (queuing buffer, table update).
2. Input rates exceed output rate (queuing phenomenon).
3. Insufficient buffers in interface units.
4. Unusually high error rates (retransmissions, etc.).

The problem is that if one condition exists, the others will ultimately also occur.

To alleviate these problems requires the use of congestion control procedures. This is the fix once it happens, to gain control. The main techniques used in wide area networks and in LANs are preallocation of buffers, packet discarding, and choke packets.

Preallocation of buffers means exactly what its name implies. That is, the sending node issues a request to the receiving node to reserve buffer space for the incoming packet(s). This type of mechanism is used presently in systems I am familiar with. In particular, an example is found in the DSDB design [Fortier 1983]. In this scheme a control bus is used to preallocate space for high-priority messages to guarantee assets will be available for their use. Other LANs have also adopted this scheme for high-interest items.

Packet discarding is a very simple method of congestion control. It operates by merely discarding any messages it does not have room for. This will work as long as the network is not too heavily loaded and the retry mechanisms on messages are constructed so that they do not feed the congestion, but help alleviate it. An example of this type of scheme is a LAN that discarded messages when it was full and a retry algorithm that would perform some form of binary backoff to not cause further congestion (Ethernet, for example).

The final form of congestion control is choke packets. This scheme utilizes explicit protocol message interaction between the interfacing nodes to control congestion. It operates as follows:

If a node reaches its allowable threshold from a source, it will send a reduction message to the source commanding it to reduce its load by some percentage. Likewise, if it has spare capacity it may tell the source to increase its flow. This is an adaptive scheme that works well, though it adds more traffic in terms of control messages onto the network, which in themselves can cause further congestion.

Flow control, on the other hand, deals with the prevention of congestion or the fixing of it once it occurs. Flow control is a network-wide end-to-end phenomenon. It has two major classes of solutions, a centralized and a distributed one.

The centralized technique deals with collection of statistics in terms of buffer availability, message throughput, and arrival rates. These are then used to compute message flow assignments for each node in the network. This in turn is used by each node as part of its routing tables (address maps) to control the volume of information allowable from source to destinations in the network.

Distributed flow control, on the other hand, comes in more flavors and deals with solving a different form of the problem. The major types of schemes are adaptive routing, isarithmic flow control, and end-to-end flow control; each addresses a different aspect of the condition.

In adaptive routing each node selects the adjacent forwarding node that will minimize delay. In a LAN with a shared media, this technique cannot be applied, as we have only one way to send out the information.

Isarithmic flow control, on the other hand, is a means by which we fix the allowable number of messages/packets that can be in the network in a period of time. In a LAN this can be accomplished by allocating the LAN assets as slices of time, and allowing a single user at a time to use the slice. This will provide a means to limit how much data is flowing totally in the network(s) in a period of time. Other means exist to perform this, such as capability access where a node must acquire a network capability token before it can attempt to send a message. This will cause possibly excessive delays if the capabilities are not managed fairly.

End-to-end flow control looks at the problem as a sender/receiver problem. This scheme puts the burden of control on the sender and receiver to limit their message interactions over the network. A means to do this is to regulate the window size (allowable number of packets that can be sent) to some agreeable value that will not cause the sender or receiver to choke up. This is a scheme found in many systems to regulate the volume of information flowing between cooperating devices in LANs today and in their wide area network predecessors.

In most simple LAN topologies the need for routing, congestion, and flow control are minimized by the topology itself and the control protocols used for allocating the media. In wide area networks, separate algorithms and processes performed these functions, whereas in the LAN environment they are an integral part of the control protocol and, therefore, are hidden from view as an explicit entity. The need still exists for these concepts, though it is lessened by the simplicity of most typical LANs.

ADDRESSING AND ROUTING

Addressing, in local area networks, typically follow one of the two well-known schemes [Shoch 1978], hierarchical and nonhierarchical.

Each of these provides a mechanism for establishing unique naming conventions for the LAN.

In the nonhierarchical or flat addressing scheme, names have no particular relationship to geography (host location) or any other hierarchy. The names (addresses) must be guaranteed unique within one single (flat) address space. Thus, this implies a more complex distributed number-generation mechanism for generating the address, or a centrally controlled facility. This method of address allocation is inherently difficult for dynamically created processes at various hosts. There are no known simple ways to accomplish this task without excessive interaction and delay at the hosts.

Hierarchical addressing schemes utilize a hierarchy of addresses that when taken together provide a unique systemwide address. Each process in such a scheme could, at the individual local level, have the same local identifier, but the global address will differ due to incorporation of additional address information for host or network. This type of scheme is analogous to the telephone network numbering system. For example, the number 1-401-792-2701 represents a unique address in the telephone company: 1 indicates country U.S. and Canada; 401 is the area code, in this case, for Rhode Island; 792 is the telephone office, in this case, for Kingston; and the last number is the subscriber address, in this case, the Computer Science Department at the University of Rhode Island.

Likewise in a distributed collection of computers forming a local area network or networks we could have addresses defined as follows:

<Network_ID><Host id><Local ID>

where Network ID indicates which local area network we are addressing, Host ID is the specific processor that the wanted process resides on, and Local ID is the specific, unique local identifier assigned to this process at its creation. Using this hierarchy we can guarantee the uniqueness of each and every process (static or dynamic) within the system under question.

Hierarchical addressing has advantages over nonhierarchical for routing and growth.

Other considerations for addressing in local area networks deal with the use of ports or clusters of processes. It is typical in LANs that I have been involved with to group resource servers and other systems service processes or user processes and to associate these with a single port address. This type of association allows a single manager process or one of the cooperating processes to access the port to get the enclosed information.

This provides a level of protection or information hiding in the system, simplifying the use of the environment. A similar construct is the mailbox in many systems. The mailbox is a repository for messages addressed to processes associated with the given mailbox. Again, this provides the hiding of underlying details or protection of the underlying processes from direct access.

In either case, the addresses are different; users in the system access via addresses such as

<network_ID><Host_ID><port_ID>

or

<network_ID><Host_ID><Mailbox_ID>

It then becomes the job of either a port or mailbox manager (part of the operating system) or the processes, to provide policies and mechanisms to map and bind the port/mailbox names (addresses) to the unique user/process names(addresses). Beyond these basics, the LAN environment must provide a means to map/bind user "logical" names to the unique system identifiers utilized in the LAN. To accomplish this requires protocols/mechanisms that perform at various levels based on the actual run-time conditions at hand. For example, if a process is loaded and put into service, and will not instantiate any new subprocesses, then the logical names can be fixed via mappings (table or code insertions) at the time of loading into the active state. Alternatively, if the process involves other processes at run time, then we need a means to provide dynamic binding at run time. In other words, there must be a mechanism (operating system service) that can provide a unique name upon demand. Typical of this is the create process command in process-based systems.

ADDRESS NAMES—BINDING AND MANAGEMENT

Introduction

In a LAN, as in any system when a process/object/procedure is created, or wants to converse with some other process/object/procedure, it must possess a name or identifier through which it can be recognized. In the LAN this name is used in determining what the process/object/procedure is and where it is located. The selection of a name can be done in various ways in a LAN environment. For instance, the user name given to a process can be used in conjunction with a user ID and system node name to give a unique, systemwide name. Another alternative is to have a global name manager that provides a unique name to a new process when asked.

Typically in systems, names exist at two levels, the user level and the system level. This implies that there are two distinct name spaces. At the upper level (human), there are typically global symbolic names. Symbolic names are those that are more convenient for human users, and therefore they have meaning to the users. For example, we may have a process identified as printer or disk server, mailbox, network I/O, etc. This name space is supported by a catalog that provides the mappings (bindings) between the symbolic names that people use and the Global Unique Identifiers that the system uses to actually access the objects.

The Global Unique Identifiers reside in a "catalog" that defines the actual parameters necessary to acquire and access an object (bindings to storage, etc.). Binding protocols, or levels, are used to define in a logical and physical way, how the namings extend down into physical realizations in a hierarchical fashion. More will be said on this in later sections.

Naming

The problem addressed here is that of how to find out a server or servers' identity or name in a system so that it can be used to perform its job. As was said previously, this is the problem of naming and name management. To address this, we will examine the techniques applied to some existing LAN operating systems. In particular we will look at 3 well-known examples from the research world: Cronus, Shoshin, and Mike. The techniques used in them are indicative of most commercially available mechanisms.

In Cronus, names exist at two levels—a relatively high symbolic-name level, and a relatively low global-unique level.

At the highest level, Cronus allows users to utilize symbolic names that provide for convenient use to humans. These symbolic names are managed by a Cronus catalog manager. This process maintains a mapping of symbolic name to global unique identifier (name).

The lower level utilizes unique identifiers at the systems level to provide a means to differentiate all objects in the system. These in turn are formed into UID tables that describe the actual physical realization of the name's object (i.e., its bindings).

In Mike there is a hierarchical scheme for names. At the global level only the "administrators" (guardian) names are known. This provides for a level of abstraction and information hiding. The naming in Mike is a logical name that does not have any connotation to its physical location.

The guardians on Mike have a unique systemwide name, while processes encased within these guardians exist and have names only at the local level.

The names of the guardians are formed into a nonhierarchical set, with their subparts named in relation to them. For example, (stack, push, si, temp) implies that the stack guardian is to push item temp onto stack si. As in the previous example, Mike uses a two-tiered approach. That is, it possesses human-oriented names and machine-oriented names.

In Mike the global unique identifiers are built as the concatenation of a node identifier and local unique guardian name. Those taken together provide a global unique identifier.

The Shoshin testbed system utilizes a similar naming mechanism. It utilizes a hierarchy of names to uniquely define items in the system. The breakdown is shown as service class name, instance name @ host name. This provides a way to uniquely address any service, instance of a service, and location of this service in the system. As in the other cases, it will require a mapping mechanism to actually perform the binding, although in this case bindings can be done dynamically versus statistically. Details are in Tokuda [1983].

The V system implements several levels of names. At the lowest level, each process has a process identifier. Next, each service process has a symbolic name, such as LPR (line printer), DSK (disk), etc.

The symbolic names are mapped via a kernel table to the PIDS. Again as in the previous examples, the names are associated as a symbolic (human) version and a system version (PID) [Cheriton 19XX].

Bindings

The binding of names to physical realizations is very critical and often overlooked. The binding can be done at process definition, process compilation, process loading, or process runtime. Each has unique aspects that restrict users to what can or cannot be accomplished with their entity.

Binding performed at definition time has been done with the past. This implies that when the designer is coding the process, a unique ID is provided. This in turn is used to map the process to its identifier for use in the future. This type of binding is extremely limiting and not used in any reasonable system.

Binding at compile time is an option that has been used in past systems. It required that the compiler or compiler manager know the current mapping and could allocate a UID for this process. A more useful means is to do mappings at load time. This again, though, is not too flexible but does provide a means to better utilize UIDs in the system. When a process is initialized, it is given a UID to have and use until it dies. When it exits, the UID is returned to a pool to be reused. This is all right if no other process will use the UID directly to address a process, else when a new process is initialized using this UID, erroneous control flows and processing will occur.

Finally, bindings can be done at run time. This implies a dynamic binding mechanism that will cause the given name to be bound to a UID only during run time. And even more, it could change during run time. Examples of these binding classifications can be found in Tsay [1981], Tokuda [1983], and Scha [1983].

INTER/INTRA NETWORK SOFTWARE

With the proliferation of personal computers and intelligent workstations, an attendant explosion in the number of computer networks has occurred. The problem with this plethora of systems is the need for communication among them. The variety of LANs developed and installed has been linked to wide area networks and to other LANs to produce large resource-sharing environments.

The motivations for linking these diverse networks arose directly with the increased growth in public networks, as well as the explosion of local area networks. A more important impetus for internetting was and still is intra organization connections allowing for sharing of costly resources and for synchronization of activities. The need to converse and share data also swept to inter organizations to allow connection of various assets.

These organizations wished to be connected in order to provide terminal access to time-sharing systems, for remote job entry, for bulk data transfers, for transaction processing (distributed operating systems, point of sale, reservations, data collection, funds transfer, as well as mail), and for multimedia teleconferencing for example.

The issues with multinetwork systems deal with naming, addressing, routing, and information representation. These problems are particularly difficult when networks of different designs, protocols, and administration must be interconnected.

Naming in a LAN is performed in two fashions; either as logical entities or as physical entities. When logical naming is used, the users are allowed to use any representation they wish as a name for their process or task, etc. The logical name need not be unique among users in the sense that multiple users could use the same logical name for a file. For example, if user A wishes to have a file called shakeout and user B also wishes to use the same name, logical addressing would not prevent this. That is, the system via the use of a mapping algorithm or table would be able to tell the difference. The system would take the logical user names and convert them to an internal form using possibly the user identifier concatenated with the logical name or use the UID as a key to encode the logical ID. In either case, the result is a system-unique identifier for the logical identifier supplied by the user. To work, such a scheme must provide a means to uniquely define and allocate user identifiers, and have a scheme to map or translate logical to physical IDs.

Once a unique name exists for an entity (file, process, task, etc.), the system can use this as an addressing tool. The logical ID is translated via an address translation process into a physical address.

In physical naming, the user supplies an actual address as the name versus some interim representation. The problem with this is that it does not allow for as much flexibility or growth in the system's user pool. The burden of knowledge is put on users versus the system.

The variance in use of logical versus physical names lies in the cost and flexibility. Logical addressing is more natural for users to use and provides more flexibility to users, as well as understandability of their operations. But it is not without its costs. Logical naming forces an increased cost to the system to perform the process of translation, though the translation can be done at a fairly low cost if intelligent schemes are used.

Physical naming is more restrictive, but is optional for the system's use, at a cost to users.

From the perspective of gateways or bridges between networks of heterogeneous and homogeneous structure, logical naming appears to be the best choice, since it removes from the users the burden of knowing where a particular entity is located. Again the problem is that it will force the bridge or gateway to require addressed processing/storage resources to perform the translations.

Bridge/Gateway Addressing

Addressing in a multi-LAN network requires added features over basic addressing. The LAN-to-LAN addressing typically is performed using one of the two basic mechanisms referred to earlier; that is, flat or hierarchical addressing. Flat addressing refers to one large address space mapped to the various LANS, their nodes, and their intended processes. This is a network management nightmare and should be avoided. Typical multi-LAN networks utilize a hierarchical mapping scheme that can be physical or logically represented. The preferred scheme, and the one most often seen, is hierarchical logical addresses. For example, if I'm a user on a university network, say University of Lowell in Lowell, Massachusetts, to address a user on the VAX D computer at Naval Underwater Systems Center, I would need to know the user's name, his computer's logical name, and the network he resides on. In this example, the address looks like:

User ID on VAXD Nuscnet on ARPAnet

Name@NUSC-ADA .ARPA

likewise, to get the other way back to Lowell, which is connected on multiple networks, requires other intermediaries as follows:

User Maincomputer LAN WAN NAN

Name@computer.Ulowell.EDU.ARPA

These transmissions would require the message to travel over multiple networks, through gateways and bridges, to ultimately get to the final destination.

The addressing mechanisms at each bridge or gateway may require extensive mapping data structures to translate names from one domain to another. If a "clean" hierarchical addressing scheme is used, where each address is by end-user unique ID, node ID, Net ID, or Wide Net ID, then the translation can be broken up into a hierarchical scheme where the address is broken down by wide net first, then to local net or wide net, then to host or local net, and ultimately to end-user process. This hierarchical translation can be performed much more efficiently than a straightforward total address translation at once.

Bridge/Gateway Routing

Routing, as was indicated earlier, is related to the addressing scheme utilized. Routing algorithms utilize the addresses as a means to determine where to send this message.

Routing can be done as in the ARPAnet and other wide area networks. Addresses to outside the LAN can be sent to the gateway and then sent to all other gateways it knows of. The address(data) can be sent to the gateway that maintains mapping information. If it knows this addressed unit, it sends it direct; else it sends it to known gateways that might know it. As alternatives, the gateways could use strict routing mechanisms such as flooding, and send it to everyone; or directory routing, lookup addresses and send to named addresses, or hierarchical routing, where one strips off the message header and utilizes the embedded information to determine an address.

Routing can be done as a source job. That is, the source (node) must know the addresses for all destinations it will use. It then directly, via simple store-and-forward messaging, sends it to the destination. This requires high intelligence at the sources and allows for simple dumb routers.

Routing can be done as a circuit switch, where connection tables in routers provide the means to link sources to destinations. This scheme requires shorter addresses in the packets. The routing algorithms require intelligence for decision making and for the maintenance of route tables/mapping keys.

TRANSLATION

Another issue in internetwork communications is information format and representation variances. The job of a translation process in the gateway or bridge is to translate representations from one LAN to another. The methods for this vary and at present most are left to the end-point users to perform. The intermediaries translate control information and leave other details to the end-to-end services.

Other functions of the gateways are flow and congestion control, fragmentation and reassembly, accounting and monitoring, access control criteria, basic security, and protocol translation.

SUMMARY

This chapter addressed the issues in data link and network transfers. Included in this review were concepts for routing and flow control of messages, naming and binding of messages, addressing, and bridge and gateway responsibilities. The goal was not to provide an in-depth view, but an overview of the details associated with this level. Details of such concepts can be found in Tannenbaum 80, and Stallings XX.

6

LAN SOFTWARE POTPOURRI

INTRODUCTION

Local Area Networks have been around for approximately 20 years, although their true use in integrated systems is just beginning. The reason has been that software to control this new technology and to make it useful to users has just begun to stream into the marketplace. LAN software has been developed to provide integrated operating systems, database managers, file systems, and information banks, as well as a variety of end-user applications. The door has opened for LAN software vendors to enter and sell their wares.

As shown in the previous chapters, LAN software provides coverage of a wide spectrum of a computer system's components and delivers a variety of services. LAN software is used at the low end to decipher incoming signals, to check for errors in transmission, to validate and authenticate conversations, to provide synchronization between users, to provide for virtual channels, and to provide transparent access to other users, applications, devices, and services over the LAN. At the high end, LAN software provides the services that users see or are directly knowledgeable about. For instance, LAN operating systems affect how users see their machine and others; it could provide transparent service or force undue overhead on the users to know of distribution and provide for the workarounds themselves. Database services are provided at this level to deliver access to information potentially stored throughout the network. In addition, the LAN provides system services such as error detection and correction, fault isolation, virtual terminal services, virtual file services, and management services, to name some. End applications that directly serve the users,

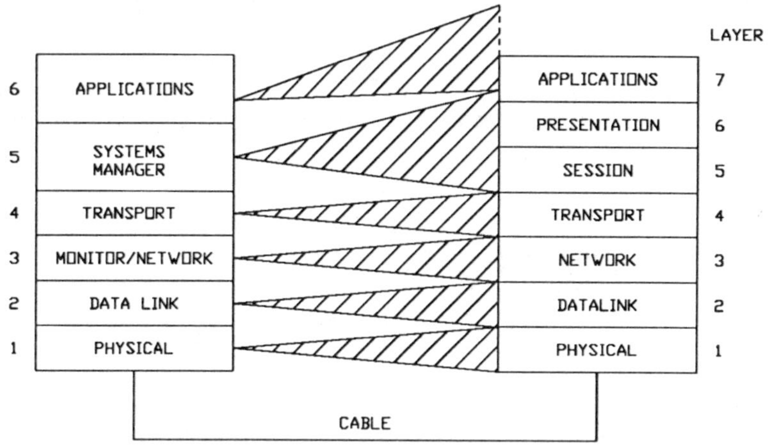

Figure 6-1. Book—ISO/OSI Reference Models.

and utilize the system's services layer include software for computer-to-computer FAX, mail, notepads, calendars, bulletin boards, print server, accounting, word processing, teleconferencing, CAD/CAM, and many more to come.

What all these have in common is that they are software packages that run on local area networks in order to provide enhanced access to information and services that could not be adequately provided within a monolithic central computer complex.

To provide a comprehensive view of all products that are available for LAN environments would be an impossible task. Therefore, the intent of the remainder of this chapter will be to provide samplings of some products available for LANs and point to places where further products and more detailed information can be found. To organize this presentation into something meaningful, I have decided to follow the layering scheme used in this book, which will provide examples of products for each layer. Figure 6-1 shows the layering mode used in this book as compared to the ISO/OSI reference model.

Figure 6-1 roughly shows the correspondence between ISO/OSI layers and the book model. Basically the four (4) lower layers map closely to the OSI model used. The difference tends to deal with the class of services and the volume of services offered by the layers. The upper two book layers diverge from the ISO/OSI model; the systems manager layer not only provides simple connection establishment and maintenance services as in the ISO model. It also addresses system maintenance, management, and operations support. The applications layer of the OSI/ISO model includes features this author feels belong in a systems management type layer versus an applications layer. Applications are viewed as services supplied to end users, such as inventory package, accounting packages, database package, teleconferencing package, etc.

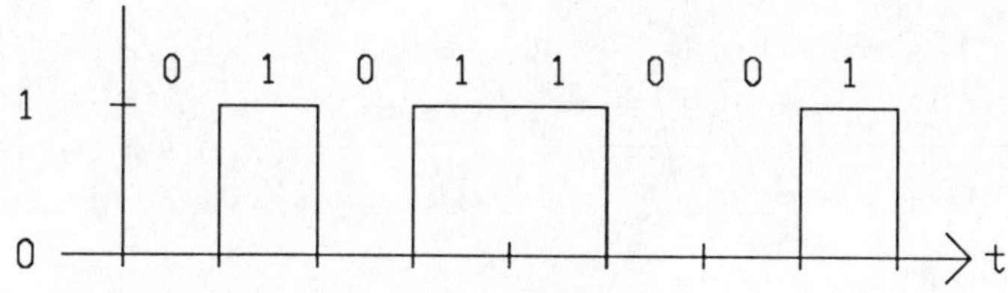

Figure 6-2. Digital Transmission of a Byte 01011001.

Physical Layer

This is an easy layer to examine from a software perspective since there isn't any software used at this layer. This layer is concerned with the physical transmission and reception of bits over media. The operations required could be partially implemented in firmware (software like code embedded in a read-only memory) that provides for interrogation and recognition of simple patterns from the bits that can then be used by upper layers in performing their functions.

Typical operations at this level include encoding of signals for transmission using techniques such as differential Manchester, Biphase L, NRZ-l, delayed modulation, Bipolar, and many more. In addition, before physical transmission, digital signals are augmented with checksums' parity or other means to provide for error detection and correction. Actual signal transmission occurs via many means. For example, it can be done strictly by digital means where the signal to be sent is sent as discrete square pulses (Figure 6-2).

This type of transmission works well for short distances or with repeaters in the circuit. Other methods of transmission include analog means such as amplitude modulation (Figure 6-3a), frequency modulation (Figure 6-3b), or phase modulation (Figure 6-3c).

In all those cases, the hardware simply transforms the digital signal to one that can be sent over analog media. Typically, this is used for communication over phone or leased lines. In any case the emphasis at this level is hardware plugs, wires or fibers, and digital electronics that work together to move bits over the network from one point to another. There is little or no software to speak of and, therefore, we will go on, leaving details for other texts, such as Stallings or Tanenbaum texts on networks.

Data-Link Layer

This layer begins to bring in some terms we are all familiar with, or may have heard of. This layer deals with establishing, maintaining, and releasing connections between

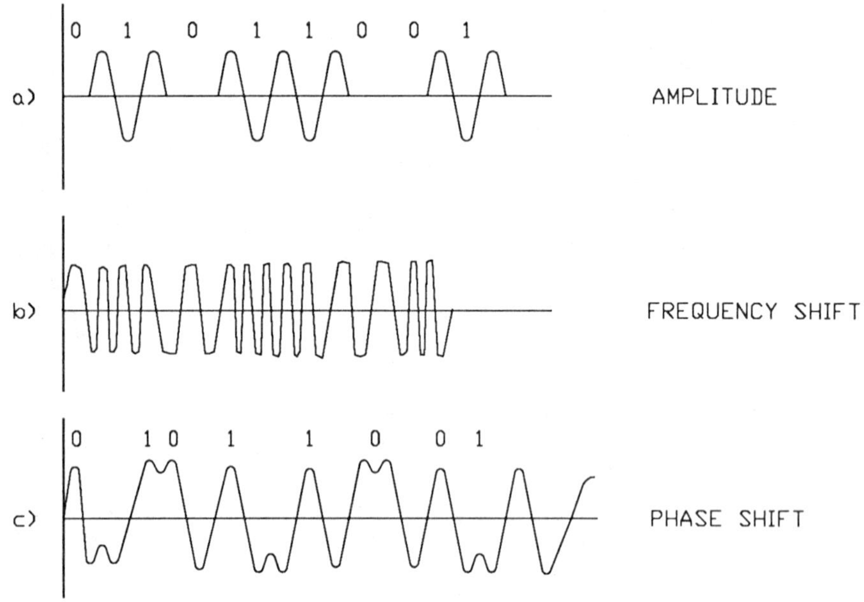

Figure 6-3. Analog Signal Transmission: (a) Amplitude Modulation, (b) Frequency Modulation, and (c) Phase Modulation.

two systems. Data link protocols provide for the formatting roles and how the formatted data is interpreted from end to end. The most familiar forms of components and protocols would be those found in the IEEE 802 standards; in particular, the 802.4, 802.5, and 802.3 standards. These standards provide mechanisms to connect devices using token passing and carrier sense multiple access schemes. The services provided to package and format data at this level can be done in hardware or in software. In most systems, it is done in hardware with many VLSI chips being developed to support these functions. Additionally, this layer is responsible for the detection and connection of errors as well as recovery schemes such as retransmissions, etc. This layer is more of an extension of the low-level physical layer. The major difference is that the physical layer deals with the smallest unit of transmission, the bit. Whereas this layer is more interested in the packet or submessage unit, in some designs this layer also controls the rate of packet reception and transfer.

Network/Monitor Layer

The function of this layer is to provide two basic types of services to the upper transport layer. These two services are network transparency and uniform transmission costs. Transparency of the network refers to the isolation of any physical or procedural idiosyncracies of the lower classes of services to the upper protocol. That is, it provides a uniform interface to transport services no matter what ordering network is utilized. Such a service is crucial when multiple subnetworks are linked together into one larger network; or if other services that plug into the standard interface are utilized, they need no adjustments to be operated within the network. The network layer also provides for a link with error rates, link availability, throughput, and delay that meet standards. Addressing capability, flow control, and routing are also providing this layer.

A typical example of a network/monitor type service is the CCIIT X.25 standard for packet-switching networks. For more details of software and products for this layer, refer to Tannenbaum, Stallings, and ISO, IEEE Standards documentation.

Transport

This layer provides the final link between systems services applications with the underlying network. Transport layer users are identifiable only by a transport address such as port or server. Details of routing, switching, and establishing the proper network connections for the user are of no concern at the transport layer. This layer simply provides the necessary controls and mechanisms required to establish two-way communications links of a desired quality for users of the service.

The transport layer operates in three phases to effect communications. These phases are establishment, transfer, and termination. During establishment, a connection is made between two transport users. This involves choosing a service class, size of protocol data units, and setup of the network layer to effect the end communication. The data transfer phase utilizes mechanisms to transfer data over the paths set up during the establishment phase. This phase may concatenate multiple messages into one to optimize transfer; larger ones may be broken into smaller ones, and errors are corrected and recovery performed. Termination phase provides the means to release the assets corrected and held during the remainder of the transport activities.

SYSTEMS MANAGEMENT

Systems management as discussed in previous chapters deals with software that provides the link between the network and applications. In addition, this is the layer where software that provides basic services for applications resides. For example, this layer provides operating systems services for getting into the network, acquiring a network address and name and for providing other basic operating systems services such as error detection/correction, systems availability management, configuration management, file and I/O control, multiple resource sharing, remote job entry, remote job invocation, message passing, synchronization, and many more.

Operating systems for LAN environments run the gamut from the simple open a link and hold it, up to full-blown environments that handle all aspects of distribution for the users. The following pages examine a few environments and provide insight into some available LAN operating system products. The products discussed will include:

NETBIOS	Banyan
Appleshare	Macserve
TOPS	IBM PC/LAN
NetWare	Star LAN
Lonet	More
Vines	MS-Net
3T	NETSTAR

These products rely mostly on host operating systems in order to run in conjunction with and to supply overall applications support. A typical environment, such as MS DOS, is a simple user operating system that provides process operation and service management for the computer. Inherently, it performs interrupt handling for I/O and process completion; disk, tape, and terminal control; file management, and error handling. Other typical base environments would be the UNIX operating system, XENIX operating systems, OS/2 operating system, VMS operating system, etc.

The focus of this section will not be on these operating systems, but instead on add-ons and enhancements to support LAN operations. These enhancements provide the local operating system with capabilities to receive requests from many applications programs at the same time and satisfy them by allocation of networkwide resources. Basically, networking software must be able to recognize users, determine their rights, determine their request, and provide the correct path to the appropriate server(s) for the request. The key to making it all work is to keep it transparent from users and minimize impact on present software investments. The LAN-OS developed to date build on top of existing operating systems; rip out part of existing operating systems and replace them with new, more "parallel operating" pieces; or build from the ground up. The best choice for future growth is the last one, but it is less reachable in today's market. But systems evolution and user demands will ultimately bring it to pass. LAN operating systems have as their goals to provide management for sharing resources (printers, disks, tapes, processors), remote access internetworking, fault tolerance, security, distributed applications support, and simple/complex intermachine information transfers.

The following paragraphs will review a sampling of LAN operating systems that presently provide some of the aforementioned capabilities.

Artsoft, Inc.

Lantastic, Artsoft's network operating systems product, runs within DOS-based systems that possess NetBios software on its LAN interface cards. This operating system requires approximately 13K of memory for the kernel and approximately 30K more for extended service.

Lantastic supplies services for users to share resources within a secure environment. Security features allow trusted users to read, write, create, and delete files as well as create directories and execute files remotely. These features, along with printer spooling capabilities and remote server stations, provide basic network services to users. It also has basic mechanisms to collect statistics on operations that could then be used to construct management software capabilities.

This product is a nice, small, neat package and provides users with a relatively cheap networking capability to build upon.

Banyan Systems Inc.

Vines is Banyan's product for a network operating system. This product runs on a dedicated server and interfaces to the network components via their NetBios component. This product delivers a high-powered backend server that performs all the interfacing for user processors involved in remote accesses. It provides synchronization and control for printing, file manipulation, and communications.

Vines provides an easy means to link networked components into a "system." The management tools provided, as well as E-mail and remote access links (dial-in facility), provide extensive service to users. Vines provides logical naming mechanisms to hide the structure and topology of the LAN from the users. Like Lantastic, Vines offers a security system of sufficient power for most applications. Vines possesses the capability via its TCP/IP service to link the LAN to a variety of other networks supporting this protocol without addition of hardware.

Newer versions of Vines will provide more integrated services for user applications, such as allowing more distributed resource sharing versus strictly backend server-based services.

CBIS Inc.

Network-OS is CBIS's network operating system product. This operating system can be purchased to operate on a variety of LAN products such as IBM's token ring, 3Com Ethernet, Western Digital's Starlan, and many more. It requires the DOS single-site operating system as its host and builds on this to construct its environment using NetBios extensions. This operating system supports message passing that can be used to build advanced NOS features, as well as basic NOS functions to acquire the network and access resources such as printers, disks, and tapes over the LAN.

Overall this is a good operating system that will allow for multiple networks to be linked together.

Corvus Systems Inc.

Corvus sells a product called Omninet PC/NOS. As in many of the other cases, PC/NOS operates within a DOS environment and provides very basic network operating system services. It provides for print spooling to a print server, disk control, and file management from a data server, and simple message-passing capabilities. From the management side, it has mechanisms built in to allow users to examine the status of queues and other elements the network. Security is maintained by use of a password and logical name scheme as well as through authorization lists that provide control of access to network resources.

This is a network operating system built for a special hardware package, that of Corvus, and provides good basic services within this environment.

Easynet Systems Inc.

Easynet is a bundled system in that you buy the hardware and the network software as a package. The package provides an excellent set of utility programs for user systems management and, as in the other cases, provides the basics for file servers and other resources servers. The drawback is that you must use their LAN with the operating system, which makes you locked into their equipment. It is simple to operate though slow, from benchmarks I have seen.

IBM

IBM PCLAN utilizes Microsoft's MS-NET programs as its basis. PCLAN provides users with access to remote workstations or network servers for printing, file service, tape drives, and many other resources. A message service allows users to send notes to each other or to process in real time. This is a basic system with no flags and whistles. If you need other capabilities, you either build them on top yourself or find a third-party vendor.

Overall, it's a simple, inexpensive way to get a LAN up and running for real work to key in.

Novell Inc.

Novell's product, called NetWare, forms the basic hub for a wide array of LAN-supported software. Novell's network operating system provides users with a fault-tolerant environment under which they can protect and manage their network software and files. As in the other operating systems, NetWare provides access to and use of remote assets, including print servers, file servers, and tape servers. It provides for message transfers between user nodes and electronic mail as a basic feature. Auto logging of activities assists in network management as well as with recovery. A feature of NetWare servers is that they can be internally configured to link to more than one network and more than one type of network. This allows your systems to share data and services more effectively.

Novell's Remote Procedure Call capability (RPC) provides the ability to construct any type of distributed applications that require a client/server type model, where one calls the other, goes on, waits, or whatever for the callee to complete. The called process can respond back and synchronize actions, etc. Such a capability provides immense power to applications developers.

This network operating system is well-supported and documented, and provides many of the features needed to develop more advanced applications. Any serious LAN user should take a look at this as an option.

Sun Microsystems Corp.

Sun Microsystem's subsidiary TOPS has a product called TOPS. This hardware/software combination allows users of Macs, PCs, and Unix-based systems to be linked together into a LAN. Basic features are service modelers that allow PC users to access servers for printing and file management.

An option, though, is to use the power of Unix and develop or port other features needed. It has a menu-driven interface and provides good access to all available features. It is a good alternative if access to a variety of machines and file types is required.

10 Net Communications

10 Net is again a bundled package of hardware and software for LAN interconnection and control. The software requires DOS-based environments for the host operating environment. Included in the package are the normal printer and disk servers, though they go much farther with built-in disk caching, printer spooling, management report program, interprocess communications software, as well as electronic mail, bulletin boards, and much more.

10 net provides built-in security restrictions and privileges. This provides means to protect files from users in read, write, both, or execute modes.

The features provided by this LAN and operating system allow users to develop distributed applications using the peer-to-peer message-passing communications and resource-sharing protocols available. The management component collects statistics on use and provides for a variety of display options.

Overall, this appears to be a first-class network operating system, a definite step in the right direction toward fully distributed operating systems.

3 Com Corp.

3+ Share is 3 Com's network operating system product. It, again, uses MS DOS as its core with server software MS-Net running on the servers. 3+ Share provides the mechanisms for users to access remote resources and use them from dedicated servers. 3+ Share consists of modules to manage processes running on the servers, to manage concurrent access to servers, to service print requests, file requests, and to locate users and files. In addition, using their message-passing capability via E-mail, one can construct distributed processes using the message-based DOS design, Paradigm [Fortier 1986].

Security is managed using access rights and passwords and, via the servers, control over directories provides hiding of subdirectories from users.

A weak point appears to be in the LAN management's ability to provide adequate logging and statistics. Although this deficiency is to be corrected by 3 Com, overall this network provides many features needed to build upon to reach the promised LAN of fully distributed oeprating systems.

Torus Systems, Inc.

Torus markets the tapestry network operating systems for DOS-based local area networks. This network operating system utilizes an icon-based interface to link users and resources within the LAN environment. Provided are an excellent electronic mail system with a notification component to tell the user when mail is received (nice if you wish to build an application needing some low-level form of synchronization), and the basic servers for printers, disks, and other shared resources; but it goes beyond basic when providing security. Files can have five levels of read/write protection and a password environment. The icons make it an easy-to-use environment, but at times this can be a deficiency when one needs to look at things in another light (i.e., with full DOS features available). Overall, this is a nice network operating system for the end user who will do no more, but can be a roadblock for those who want a basic facility to build upon.

Univation

Univation's Lifenet is a network operating system for DOS-based LANs. It provides a fault-tolerant environment via its ability to recover from failure conditions. LifeNet uses a logging utility to track each transaction and, therefore, provides the ability to reconstruct state upon failure. It provides a full management capability to track users' use of servers and machines, as well as network communications. The manager can interrogate and adjust usage attributes as the system operates. On the security side, LifeNet provides mechanisms to protect files by password or by hiding them from use and limiting access by assigned privilege.

On the basic feature side, LifeNet offers users the capability to share file servers and database management services, to share printers, and to send and receive communication from users. Using this feature, one could construct a full-blown distributed application, although it may not be a "real-time" implementation.

Overall, this is a nice package and provides the basics and some for LAN software applications developers.

Western Digital Corp.

Western Digital's network operating system VIANET is an MS-DOS/Unix/Xenix-based environment. This environment provides the means for any workstation on the network to share resources with others. Using this capability, a VIANET user can spawn processes and use resources throughout the network—although at present you, the user, must coordinate your own interactions. VIANET is a basic LAN operating system providing you with the tools to construct more robust environments.

Microsoft Corp.

Microsoft Corp. markets a local area network operating system called OS/2 LAN Manager. It operates within the OS/2 or DOS environments and provides network-transparent access to users of the LAN. As in the other LAN operating systems, OS/2 LAN Manager provides mechanisms to set up server nodes and provide resource sharing across the LAN. It provides security via access privileges granted, revoked, and changed by a network administrator. Access privileges to read, write, delete, create, and execute files are provided, as well as login checking via passwords. The management tools provide for logging, analysis, and display of network attributes such as directories, files, print queues, pipes, communications queues, and other systems parameters.

A key feature of this operating system is its interprocess communications feature. This feature allows users to create named pipes (conduits) through the network so that multiple processes running on various machines can be linked together by interprocess communications. Using this capability, developers can design a wide array of distributed processing applications.

OS/2 LAN manager provides a full and uniform operating environment for users to operate within now, and to expand in the future.

Beyond these basic LAN network operating systems to link workstations and personal computers into unified environments, the large mainframe and minicomputer manufacturers also offer users a wide range of local area network solutions.

Wang Computer offers its Wangnet and WangPC-Nets to link its various minicomputers, mainframes, and PCs into unified environments. Wang offers a wide array of LAN operating system services as enhancements to its VSI operating system to provide network services and distributed applications to users.

Digital Equipment Corporation offers its Decnet solutions to deliver the capability to cluster large minicomputers and workstations and PCs into an integrated computing facility. Decnet software, in conjunction with VMS software, provides extensive capabilities to share resources, to link together processes from multiple machines into one distributed application, and to extensively share information across the multiple machines. DEC is committed in the future to providing enhanced service to users and developers of distributed applications. DEC supports a wide array of network protocols such as TCP/IP, FTP, as well as message-based communications.

Many other vendors such as AT&T, IBM, SUN, Apollo, Hewlett-Packard, and Zenith offer their own versions of network operating systems and services.

The future lies in developing added features to enchance applications development on these architectures—in particular, the use of remote procedure call capabilities, object-oriented interactions and message-passing paradigms to construct environments where highly interactive applications can be developed across sites and made to operate efficiently. These represent basic features for use in developing synchronization techniques and interprocess interactions necessary to realize the full potential of LAN architectures and capabilities. The future lies in developing more robust environments that better utilize the assets of a LAN consisting of not only PCs but terminals, workstations, minicomputers, mainframes, and supercomputers, as well as highly specialized resources such as database machines, etc.

LAN DATABASE MANAGEMENT PRODUCTS

With an operating system in place, one can begin to think of writing applications to run over the LAN. An important applications program, as well as an overall systems service, is the information manager or database manager. Database applications provide for controlled access to network wide stored information with the major emphasis in three categories: online transaction processing (like the telebank machines), distributed databases, and database servers.

Typical of a LAN installation at an enterprise is to provide controlled access to the totality of the enterprise's data banks, no matter where they are stored. For example, regional and national banks have linked themselves together into LANs and WANs to provide us with the use of our bank cards regionally, as well as nationally, to access our funds. This feat is accomplished by providing distributed/federated database management and transaction processing over the networks. The user inserts his card, the bank's database checks its validity, then searches its known space to acquire this user's file. The transaction operates on this file locally and then issues the same transaction over the net to the user's home bank. Another user of LAN technology and databases is the airline reservation system. This system uses LANs to connect terminals and multiple databases into a large, integrated database. This system provides users with access to flight, airport, and airline information throughout the country.

In both cases, as well as in many others, the LAN provides the means to link and integrate the multiple databases into a tightly grouped, distributed database, or a loosely coupled, federated database providing remote access to the stored datums. The database is a storage container in which these datums can be organized in some controlled way. In addition, it provides the mechanisms to control how these datums are accessed, stored, related, or otherwise viewed.

Database management systems provide these basic services plus many others to assist in their creation and use. A database management system should include components for database design and maintenance, database administration, data entry facilities, report generations, backup and recovery utilities, as well as ad hoc query interfaces and transaction processing facilities.

Adding a local area network complicates the design, development, and operation of the database. A database management system ported to a LAN can be fully distributed, with all control and data, as well as the DBMS functions done in a totally distributed fashion, or can simply distribute the data to multiple hosts and give users directories of where to find what. A third alternative is to not distribute the database but simply distribute the access to the server host. In reality, what is seen today is the latter two with some slow migration toward the ultimate fully distributed databases.

In the first case, the database manager is accessed by a set of remote "terminals." These terminals issue requests, and the DBMS processes them for the LAN user. The distributed database typically uses the client/server model [Fortier 1987] to control the interaction of the database and the user. The client requests service from the LAN database manager. A connection is made to an appropriate server who then provides either transaction-based access on possibly raw data transfer.

The site where the actual processing of the query is performed is dependent on the sophistication of the client and on the server. In some cases, the client formulates the query and accesses the server to get at the raw data; the client then processes the raw data to get the required answer. The server in this case is simply a file manager with little intelligence to perform database operation. In other systems the user (client) is dumb and the database server is smart. In this case, the user station simply has the capability to formulate a query; the operation on the query is handled by the server, as well as the operations on the underlying database. The server in this case is the point where all processing is performed and could become a performance bottleneck for the LAN.

A third possibility, and the more optimal, is that both client and server have intelligence to perform database operations, and the power is used. In this type of combination the processing of transactions and queries can be divided between the client and server, providing for faster service. This is the true distributed database avenue and will be the direction taken by future environments to provide enhanced service to users.

Beyond these basic configurations, the database can also come in varying flavors. Data can be managed as relations in the relational model, or as linked files in the network, or as hierarchies of linked records as in the Cobol databases. The most popular, though, from a present LAN viewpoint, is the relational model.

Beyond the data model supported, and even more important from a user perspective, is the data manipulation service. Typical is a simple query capability to read, write, create, delete, and modify information; but the future lies in online transaction processing where user requests are provided full database services (protection, etc.) bundled into transactions. Online transaction processing provides the database user with a powerful tool. The basis of the transaction is its feature of operating either totally or not at all. That is, the effect of the transaction operations on the underlying data sets is either completely performed, bringing the database from one consistent state to another, or not performed at all (it is either aborted, or rolled back to the original consistent state). The all-or-nothing feature of a transaction along with online, real-time execution makes it desirable in a database management system. Transaction processing provides these features via the use of two-phase commit protocols [Date 1984] to aid in the synchronization of operations over the multiple databases and to ensure correct operations. This, along with checks on bonding conditions and on conflicting operations from other sites, aids in ensuring that database consistency is maintained while keeping the database accesses as concurrent as possible. In addition, transaction processing must ensure security of the database via authorization checks, etc., and must provide crash recovery on failure.

These all relate down to the basic features a database management system should have. For example, recovery sounds simple enough; but to have it, the database must have online logging of transaction steps, must have the capability to store copies or versions of data and features to roll forward to a good data item if it completed before a failure, roll back to a previous good data item for those that did not complete (commit) before a failure, and restart features to reinitiate transactions that must be done automatically.

One must look to see if the database supports transaction processing, if one wants it, and what language (SQL) is supported within it. Are good database tools provided to construct data tables, records, relations, etc.? Are good application development tools provided to assist in our own applications developments? Beyond these, does it run under various programming languages or operating systems? Is special purpose hardware or software required beyond the basic database to use this system? We must also look at how it affects our run-time environment in terms of memory required, secondary storage buffer management, I/O overhead, and overall performance.

If we want a multiuser environment, what kind of locking or concurrency control is provided, what is the granularity of control of this feature, and how is overall security and integrity of the database ensured? These represent but a few concerns one must have in looking at databases for any environment. When we now insert the LAN, we add more in terms of distributed access and control related to all the previous concerns.

Having voiced these concerns, let's now turn our attention to what is available for our use. There are basically five major categories of data access managers for the local area networking environment. You notice that I said five "data access" managers, not database managers, since a few of these are more like primitives versus true database

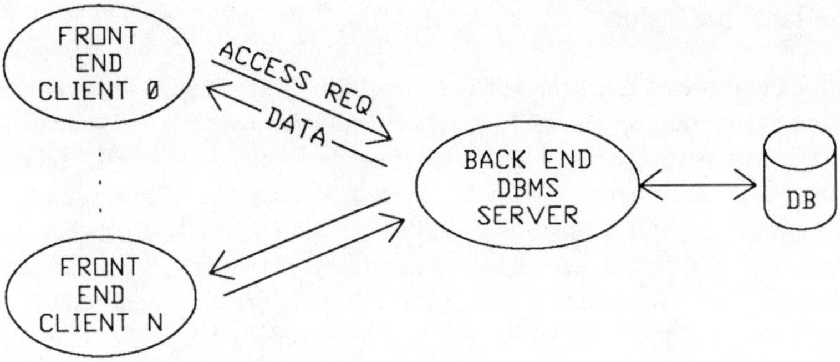

Figure 6-4. Database Client/Server Model.

managers. The five categories of server for the LAN environment are disk servers, file servers, integrated file-server based, database management, system backend servers, and distributed database managers. The first three are not true database managers, but served to lead the way toward the end goals. These provided controlled access to data stored on secondary storage devices, but did not provide any true database management capabilities, as previously discussed. The third category of server, the file-server based DBMS, provides crude DBMS features via use of remote servers linked logically to centralized DBMS.

Beyond these, true products built for the LAN environment are emerging. Integrated database servers provide unified networkwide database services to users. They provide uniform query interfaces, security, data integrity, recovery, and concurrency control in a network circle-supported package. User sites have an interface program that allows them to query and use the remote database manager node as if it were locally residing. The shortfall on this type of architecture is that it is a backend approach where all the hard database functions are performed at one or a few sites. This could, in heavily used environments, cause a bottleneck and a slowdown of overall system operations. The full distributed database or LAN integrated database provides for data management execution at multiple sites and with integrated/synchronized execution, providing for optimized performance.

The first class of LAN database products to be highlighted are the backend servers (Figure 6-4).

Those announced to date typically use SQL as their native query language, but allow users via interface applications to use a variety of access mechanisms.

The reviewed servers are offered by Gupta Technologies (SQL Base), IBM (OS/2 Extended edition), Lotus (Lotus/DBMS), Emerald Bay (Mignet), Novell (NetWare SQL), Oracle (Oracle Server), and Sybase (SQL server). This is but a sampling of products, and users interested in a wider review must search the LAN horizon on a daily basis for products that emerge.

Gupta Technologies

Gupta Technologies of Menlo Park, California, offers a SQL-based server called SQL Base. This server runs on an OS/2 or DOS-based system with NetBios as its framework. SQL Base is composed of three main modules. The first is a SQL preprocessor that parses SQL statements, breaks them down into low-level relational database operations, and determines the most optimal execution path. The second component provides efficient means to access individual's tables or attributes within them. The third component is a virtual I/O manager that provides for management of the database's physical pages in memory. This module does the buffer management, concurrency control, failure recovery, and transaction processing for the server. Taken together, these provide the three major features of user interface, database processing, and data management for the database management system.

In addition to these backend database capabilities, Gupta also markets an applications development program that allows users to develop displays and programs under the Microsoft windows environment. Another significant feature of SQL Base is that it can be run under the DOS or OS/2 environments. In addition, it can be coupled with existing IBM mainframe databases using SQL net and SQL gateway.

This product is an interim one in the sense that it provides database management capabilities now. It is not a distributed database management system, nor does it claim to be, but it is nonetheless a nice product. Details can be found in company literature.

IBM

Big Blue has its own product for a backend SQL server. The difference is that it is not a stand-alone package, as are many of the others to be discussed. It is instead part of the operating system kernel. That is, OS/2 Extended Edition (EE) has as part of its kernel a relational database management system. IBM's contention is that integrating the relational DBMS capabilities into the kernel will provide the basic facilities to expand this server into a full-blown distributed database management system in the future. The interface to the kernel database is SQL and the functions of SQL are embedded in the kernel. IBM seems to be pointing to the OEM (original-equipment manufacturer) market to use their simple interface to build more user-friendly environments. For details of this product refer to IBM.

Lotus

Lotus, the builder of 1-2-3, has their own version of a backend database server product called Lotus/DBMS. This product again services the SQL market, although it hides many of the messy details from the users with integrated menu and graphics routines.

Lotus provides a presentation manager that allows users of Lotus/DBMS to design, change, and utilize databases without reverting to procedural programming (conventional SQL). Instead, their interface provides graphical means to construct formats for a database relation, to query this database, and to use extracted data in highly graphical forms. In addition, there are means to lower into the procedural arena for writing your own interface packages. For Lotus 1-2-3/6 users, this product offers a nice way to get relational database services.

Migent

Migent has developed a backend database server called Emerald Bay. This server is built as two elements, a server element and a user element. The user or client element provides the user workstation with a local database, controlled by the local client process, and access to the backend data base server. This type of environment will provide faster access to user locally stored data and provide management of the data for the user. The server component runs on the backend database machine and provides procedural database management access and management to the server data base. Migent at present does not support any particular interface, but provides basic services for transaction processing, including logging and recovery. To get details of this product you must direct inquiries to the vendor, since not much has yet been published regarding this product.

Novell

Novell has built a SQL server (NetWare SQL) on top of its file server running on NetWare. NetWare SQL runs only under the NetWare operating system and provides backend database services to user programs and applications written using the XQL developers kit. This product is strictly a server product and is aimed at OEMs who would build applications to run on it. The product supports transaction abort and rollback on failure or crash. Vendor data provides a further description of this product.

Oracle

Oracle's server is a product developed to manage relational databases on a variety of machines and operating systems. This product supports a SQL interface and provides all the major database functions. The product provides buffer management, recovery, logging of transactions, lock management, and online transaction phases.

User client processes need only to access the server via SQL commands. The server provides all processing and control of the access for the user client process. Further details on Oracle will be given in the following section review.

Sybase

Sybase, in conjunction with Microsoft and Ashton-Tate, developed SQL server. This relational database backend server runs within the OS/2-based LANs. The interface to the server is via SQL commands, which can be embedded in user dBASE, C, and COBOL programs or be generated online. The server takes the commands from multiple users and performs the transaction processing to provide database integrity and concurrency control among the multiple users.

XDB SYSTEMS

This quick overview provides just a glimpse of some available products; many more become available with each passing week. LAN DB servers will continue to provide a service where multiple machines require access to centralized or federated database assets, although the future lies in full-blown LAN-wide distribution database capabilities.

Systems that will provide full capabilities are coming and will be here soon. The following review looks at a few systems presently available that provide more integrated database services—backend database servers.

Not many systems have been specifically built for the LAN environment, although many new products are expected in the coming years. Some representative products include Empress-32 from Empress Software; Informix from Informix Software; Oracle from Oracle, Inc.; Helix VMX from Odesta; and Ingres from Relational Technology.

Empress-32

Developed originally under Unix, Empress-32 runs on a variety of operating systems environments such as VAX/VMS and MS-DOS systems.

This product views the database as relations, accessing and manipulating them using the SQL language. The product uses a database administration subcomponent using data dictionaries to control access and guarantee the integrity of the overall database. Transaction processing is supported in both embedded (languages such as C, COBOL, and Fortran are supported) and ad hoc (straight online SQL calls), and it supports a powerful tool available to provide online interactive support to build screen displays, query interfaces, and reports.

Empress-32 databases can be centralized or distributed, depending on the physical structure of the system and stored relations. Empress provides mechanisms via views to hide the physical distribution from users, allowing for transparent use. Users can also query databases by logical name, providing for location transparency but not database transparency. Users know in this format they are accessing a separated (maybe remote) database other than that which they are logically allocated to.

Beyond these features, Empress provides for full recovery from failures using the two-phase commit protocol, and extensive logging and archival capabilities in support of this recovery mechanism. In short, this is a nice, full-performance database system offering extensive features to users, and based on Empress software's track record, they will provide future enchancements to meet some of the harder distributed database problems such as update synchronization and concurrency control.

Informix

This relational database management system was developed for the Unix environment, and utilizes the standard SQL command language as its query language. As in the previous system, it supports embedded operation with high-level programming languages. In particular, it supports C, COBOL, and Ada. It also provides an ad hoc query capability so that users can interactively query the database using SQL.

Like Empress, they provide DBM mechanisms to limit access and control how users utilize the database. Also as in Empress, this database management system provides the fourth-generation language (4GL) as an applications development support environment.

This database management system uses the client/server model [Fortier 1986] as its means to build distributed services for users. Clients reside on user machines with a variety of backend servers distributed over the network. The DBMS provides mechanisms to map the distributed data over the system using dictionaries/directories to make the database look like a single monolithic database from any site. Users can access databases distributed throughout the system, either directly (by name) or through the logical views that integrate them.

Provided in the package is active concurrency control achieved via locks and recovery through transaction logging and rollback. As in high-performance centralized database management systems, Informix maximizes its use of I/O and memory. In addition, Informix load-shares processing to network workstations to free up processing on the backends, thereby potentially increasing the level of concurrently executing users that can be supported. With its wide range of support tools and links to other products, Informix is a fine asset to possess within supported environments.

Oracle

Oracle is one of the more mature DBMS's being reviewed in this section. Developed around the relational database model and using SQL as the method for query processing and database administration, Oracle has the only ANSI-validated SQL interface. Oracle supports both interactive and embedded SQL queries running within Unix, DOS, OS/2, Ultrix, and Xenix operating systems environments. To enhance the basic SQL feature, Oracle also offers a wide range of support tools such as a report writer, math package, page layout formatters, and the 4GL environment to allow construction of user applications environments using C, COBOL, Fortran, Pascal, PL/I or Ada as the language of development.

As in many of the other cases, Oracle is built around a client server model with both being smart. This feature provides Oracle with extreme power to issue queries and perform their operation in a variety of ways. Transparency of access is maintained by the use of directories and use of these by the servers/clients. The clients maintain global schemas that define how to access data and where, although all actual data manipulations (security, updating, recovery) are handled by the servers. As in the other architectures, Oracle uses transaction logging and checkpointing to provide recovery from failures, all invisible to the users.

Oracle is a mature DBMS and should be considered a viable candidate by any serious local area network DBMS user.

Odesta

Odesta has a product called Helix VMX that provides an Iconic view of stored datums. The database is built as strings of icons with relationships built by linkages between icons. The entire system is based on visual programming where query processing, report generation, screen form design, and applications programming all rely on icon movement and manipulation. This environment is Macintosh-like and comes from this background. It supports distribution via the VAX VMS virtual file-management system.

The use of icons as the basis for the data is a novel approach and as this product matures, we may see a more usable product.

Ingres

Relational Technology's Ingres, developed first as a university research vehicle and then recoded into a high-performance product, is the last product we will look at cursorily. This product is a relational database management system; it uses the SQL interface, as do many of the other DBMS's reviewed. It provides extensive services for embedded and ad hoc transactions, including security checking and authorization using checking via boundary checking variables, concurring control using locking, recovery using logging rollback/forward. Distributed Ingres has been around and deliveries expect to be seen in this calendar year. Ingres has gone through much testing and evaluation as a single-user DBMS and it is expected that the distributed version will deliver also.

For greater in-depth views of the various DBMS seen here and for many others, refer to vendor literature.

LAN APPLICATIONS SOFTWARE

Beyond the basic software needed for point-to-point, broadcast communications and media/data management lie the real applications. LAN applications software is built on top of and utilizes the mechanisms provided by the LAN operating system and database system. LAN applications come in many flavors, as indicated previously. Applications exist for design aid, business data processing, control, and many more. The following sections will begin to survey some products in each of the previously described applications areas.

INFORMATION ACCESS

One of the first applications areas being developed is that involved with information control and distribution. This applications area has had the greatest effect on overall LAN software demands. As users have taken to getting E-mail messages, keeping their calendars on line, sending and receiving documents on line, and accessing a wide array of information stored over a LAN, they have begun to demand even greater integration, access, and diversity of applications over a LAN to perform their tasks.

The typical office today utilizes electronically transferred information in a variety of ways. Personal computers, video teleconferencing, electronic mail, online FAXes, and electronic bulletin boards are all becoming indispensable parts of an enterprise's tools in performing their trade or business.

Information transfer software such as E-mail, bulletin boards, calendars, FAX, file transfer, and teleconferencing will be addressed.

E-Mail

Electronic mail, or E-mail for short, is an applicaton designed to allow users to transmit messages among one another over a LAN electronically. E-mail has been around and used for many years within the minicomputer/mainframe communities, and has grown and found a home in the LAN community. Today most vendors supply some form of E-mail as part of their systems management software, or as a stand-alone or integrated office automation package. Examples of E-mail packages are CC:mail LAN package, from PCC Systems; Notice/PC, from CSC; Network Couriers, from Consumers Software; Document Exchange from AT&T; Higgins Mail, from Conetic Systems; Cross/Point, from Cross Information Company; PC/Orion, from Orion Software, Inc.; Sunalis, from Sun Microsystems; Mailbridge, from SoftSwitch, Inc; and ETC/EM, from Adrive.

This is but a brief listing of some products; there are as many E-mail systems as there are systems, though they are typically all built around one of the message-interchange standards such as X.25, X.400, SWA, DNA, VS, and other communications architectures.

CC:Mail The product from PCC Systems, Inc., is a full-fledged mailer. It provides features to send and receive messages, create/delete messages, append data files, text files, binary files, and image files to messages. Messages are sent to mailboxes and can be shared to provide a "broadcast" type capability. Users can view and organize mail files in any way they want by utilizing packages. In addition, this is a very easy package to master using its nice, menu-driven approach.

Notice/PC This messaging system is marketed by CSC. It offers the PC user the ability to create/edit messages; join spreadsheets, word processing files, and data files to the messages; and transfer these over CSC's Infonet worldwide network. This represents an E-mail package used to bundle information for transfer over LANs and WANs.

Network Courier Consumers Software's product, Network Courier, is an E-mail package with many features. You can create, store, read, send, and delete messages. A message can be either a simple letter, or you can attach other text files to it. Utility services allow you to strip out text files from messages and sort mail by a variety of classes. This product in addition has mechanisms to allow it to link to other networks and mail systems, and to send and receive mail from these. Overall, it's a nice, full-featured application.

Document Exchange This product from AT&T provides users basic mail creation, editing, sending, receiving, and storage capabilities. It provides mechanisms to link a variety of computer systems together and automatically translates documents and mail messages into appropriate formats.

Higgins Conetic Systems markets the Higgins Office Support package. A part of this package is an E-mail module. This module provides users with a full-function E-mail system consisting of editing routines to create, delete, modify, and sort mail messages, as well as functions to examine mail or package it in various forms. One can send mail to one person, multiple persons, or groups of people, or broadcast to all on the LAN. The product has a built-in prompt to tell users when they have mail or to tell them that a message sent was received. Security is built in so others cannot access mail not addressed to them. When this mail facility is taken in the context of the full Higgins system, one has a very powerful office management tool.

Cross/Point This product from Cross Information Company provides users with the capability to exchange messages and perform mailing using Hypertext features. As in the other cases, users can create, edit, delete, send, and receive messages, with capabilities for special character transmission.

PC/Orion Orion Software Inc. markets a complete E-mail service that provides users with the full capability to edit, create, delete, store, and send messages integrated with files anywhere on the LAN. In addition, it provides for these services with PC, VM, and System/3X sites.

Sun Alis Sun Alis by Sun Micro Systems is more than a simple E-mail system. It is a powerful office productivity package that provides tools for document/mail preparation, spreadsheet analysis, graphics production, lists management, calendar, and many more. The E-mail component is simple to use and provides users with the capability to send and receive messages and to integrate these into a variety of formats.

ETC/EM ADP's ETC/EM is an electronic mail product that provides full capability to send and receive a variety of mail types. This product provides basic features to create, edit, delete, store, and manipulate mail messages and, in addition, provides for group or individual transmission services.

Office Productivity Packages

Beyond the basics of E-mail are fully enhanced office productivity software tools. Typical capabilities included in such packages would be: an E-mail function, calendars, schedulers, timers, built-in calculators, spreadsheet programs, word processors, dictionaries, spell checkers, directories, mailing lists, bulletin boards, accounting packages, and many more potential capabilities. Examples of such tools are Word Perfect's Executive and Library, Symantec's Q&A, Conetic's Higgins, Spectre's Command Performance, Action's Coordinator, FutureSoft's Right Hand Man, and R&R's Shoebox.

Executive Word Perfect's Executive and Library packages both provide excellent office productivity enhancement capabilities. Word Perfect Executive provides a full-featured word processor capable of developing professional-looking documents with text (Vandus Fonts available) and graphics integrated. It supports a spreadsheet capability that possesses functions for financial, arithmetic and logical operations. It also possesses the capability to produce graphics in bold, underlined features, as well as search, sort, copy, and cut and paste capabilities. Executive provides an appointment calendar; an online calculator with built-in functions; a note card organizer for action items, doodlings, and any other notes; and a directory for phone numbers, names, and company addresses is at your fingertips. The library program has all these plus a shell program that allows users to integrate their own productivity-enhancing applications with those already available, and features capability for fast switching between functions, similar to how the user would work on multiple tasks on his or her desk. Taken together, Word Perfect has provided a powerful office productivity tool with these products.

Symantec Corp. Symantec's product, Q&A, provides the framework and tools for a full-service office productivity enhancement tool. This package offers functions to create and manage database files in simple English terms; to construct documents with a powerful, easy-to-learn word processor; to send and receive mail and merge files for mass mailings; to construct and maintain calendars, appointments, bulletin boards; and to integrate outputs from as well as manage operation of dBASE, Lotus1-2-3, and from a variety of other OS/2 and DOS-supported products. Overall, this product forms the basis for an extremely powerful system.

Conetic Systems, Inc. Higgins is Conetic Systems' office productivity enhancement product. This product provides functions to manage scheduling of people and resources, to maintain mailing/telephone directories, control and track expenses, provide E-mail, develop and manage a variety of documents, and keep appointment lists for all logged users.

Higgins is built around a relational database called C/Base with all data being integrated through this database.

Higgins scheduling module provides a LAN user's enterprise with an effective means to produce and maintain calendars of and coordination for user's appointments and meeting log information. Alarms and displays can be automatically generated to tell the user that a meeting is scheduled or an event is pending. Higgins scheduler provides users with places to allocate resources to meetings and maintain the time status of items.

Higgins directory provides the means to maintain files on any entity such as mailing lists, clients, etc. The E-mail system, as described earlier, provides the means for Higgins to share information among network users.

A feature of Higgins that helps organize people is shared to do lists and appointments. People can check each other's calendars and coordinate extras easily. Supervisors can check "to do" lists, fill in assignments, and see results of all employees without leaving their desks.

Higgins provides extensive reporting modules to assist in accounting and expense tracking for an organization. Beyond these, Higgins has the standard features such as a calculator, clock, doodler file, and a word processor. In short, Higgins is a full-service office organizer, and should be examined by any enterprise looking to the office LAN support software arena.

Spectre Software Spectre offers Command Performance as their product for office productivity enhancement. This package offers to the LAN user the standard word processing software as well as an online calendar and bulletin board type feature, an appointment scheduler that has some built-in features to cross check other's appointments to coordinate meetings, and other nice features. It offers a basic menu feature with the ability to jump between Command Performance modules and other displayed modules. Users can customize their menus and build-in security on their controlled files. This is a nice, simple program and offers some of the basic features wanted in an office productivity tool.

Action Technologies, Inc. Coordinator is Action Technologies' product for LAN office management. This product was designed as a work-group integration support environment. To provide this integration, Coordinator possesses three managers: the conversation manager, schedule manager, and connectivity manager. These each provide services during different phases of a work group's interaction. The conversation manager is the glue of the product. It provides the basic mechanism to produce documentation, deliver it to named users, and check status of ongoing activities. In addition, it keeps a chronology of all proposals, requests, commitments, and agreements made between interacting users based on their distinguishing data.

The schedule manager provides LAN users with calendars and appointment schedules for daily and future activities. Scheduling can be done as a user-interactive activity and can be derived by the software based on conversation decisions.

The connectivity manager provides functions like E-mail and voice over phone lines. The E-mail package has built-in features to let you know when you have mail and how urgent it is to read now, for example. This system provides a node group productivity platform, but may lack some of the flexibility to grow, as was seen in some of the other packages.

Future Soft, Inc. Right Hand Man, Futuresoft's LAN office productivity software product, is a nicely integrated package of utilities for aiding LAN and user productivity. This package provides fast access to its toolset components such as E-mail, calendar, directories, calculators, editor, notepad, typewriter, and import/export window.

The E-mail system allows you to create, edit, and send mail as well as create mailing lists to send multiple messages to. The calculations are similar to those seen on other machines and provides all the basic functions and some extras. The directory system allows LAN users to create an extensive telephone book with over 12K entries. It offers support to manipulate directory records to search, sort, access, and recover them.

The editor is full-featured and allows manipulation of both text and graphics. Notepads provide for doodling space as well as a bulletin board-type capability via the LAN-shared notepad.

Appointments can be scheduled and coordinated over the LAN and alarms set up to tell you of impending actions.

While in Right Hand Man, users can send and receive data via the export/import capabilities and run other DOS programs through the macro routes available. This is a very capable LAN office tool and is reasonably priced.

R&R Associates Inc. Shoebox 3 is R&R's entry. This product is not as full-blown as the other office productivity enhancement products, but it does have some nice features. It provides mechanisms to manage time, using three packages. One is for "To Do's," one for meeting scheduling, and one for expense tracking. From this perspective, it provides more management tools than productivity tools, although what it offers is good. The "To Do's" package gives you a nice, easy-to-use package to set up daily worksheets and track performance.

Meeting Maker gives you a means to schedule meetings and coordinate the same with all other connected individuals. This tool gives you a powerful means to coordinate the timing of meetings. The expense tracker provides easy-to-use menus to walk you through the collection and management of day-to-day business expenses.

MISCELLANEOUS OFFICE TOOLS

Not all packages are created equal, and many of them offer some subset of full-featured systems, although as time goes on more will evolve into full-blown environments. Some tools provide simple group document development, while others provide checking for the same or editing, and other packages provide for resource sharing and integration.

Framework Software

Under this category of miscellaneous office tools, we find packages that allow us to configure how we as users will view our environment. In other words, they provide a way for each user to customize their LAN network environment's interface around the user's own preferences. They don't really add much to the overall LAN except overhead, but they can be office-enchancers if they make the computers more usable to the new computer types.

Leader System's Wizard is such a package. This product delivers a menu management program that provides a user-friendly interface to manage your online files and programs. Also included are features for tracking who is using the system and in what ways. It includes a calculator function, auto driver to support phone delay, and security based on password and access lists.

Direct Net from Delta Technology is another menu management and LAN interface program. It provides users with the means to access LAN-based services from simple menus no matter where the service is located. Direct Net manages addresses and linkages to remote servers for users, hiding network intricacies from the users. As in Wizard, this package provides means to collect statistics on users, but with no built-in features for authorizing who they are. It simply lets them in, takes their name, and tracks what they use.

Le Menu from Bartel Software Inc. is another menu management and access scheme for LAN users. This program provides users with a means to construct maps of their LAN available programs and files, build them into user menu systems, and make them available for use by LAN users. This is not a sophisticated program, but it works and provides basic features for sharing of files and programs with network users.

Deere Sof Inc.'s Magic Menu is, again, in the same family of support programs. As do the other packages, this one offers users a capability to construct menus to allow users to utilize network software invisibly. The program requires users to log into the menu subsystem and network as separate entities, although other than this, no other restrictions exist. It is a simple package that provides adequate resources to to the job.

Two other products, LANscope and Landscape offer a slightly different slant on network management. These provide services more for assisting users than building environments.

LANscope from Connect Computer, for example, provides a network global managing scheme a la 3+Share.This provides a nice means to logically group, control, and track LAN usage. It integrates user and LAN menus into one space, providing for LAN isolation. The package provides tools to construct views of the LAN and provides access to a universal database to integrate all things together.

Landscape, on the other hand, is a LAN utility to organize and manage the naming and location of files within the LAN. It's a simple program that provides LAN directory service for users, giving them the capability to locate and access files on the system.

Document Preparation

Some products such as Forcomment and Compare Rite are not part of a fully integrated environment, but instead are stand-alone packages. These provide document preparers editors and libraries with means to check differences in files (Compare Rite), or provide a LAN tool for document preparers to distribute documents and get comments back online (ForComment).

Specialized Office Support Equipment

All offices use facsimile (FAX) machines to distribute documents over leased lines on the telephone network. Such devices have been around for years and have proven their worth over and over again. LAN users are office support people, and they need FAX services.

In addition, traditional FAX machines and solutions limit the volume and diversity of transmissions possible. Enter the new work of specialized LAN servers and PC servers to provide the service. By supplying specialized I/O cards and software for your PC, one can convert any or all of a company's PCs into FAX servers. Beyond this, it opens up the door for better utilization of the FAX material than was previously available.

The new, emerging LAN products begin with a facsimile board that contains the same hardware and interface elements as required by CCITT facsimile specifications. Along with this board, which can be plugged into a PC or workstation, we could add laser printers, optical scanners, high-speed modems, and mass memory devices. These, along with software, provide for highly capable Facsimile Plus services.

The software performs basic facsimile functions such as facsimile file management, telephone number directory storage and retrieval, and conversion software to convert nonfacsimile formats to facsimile format.

Add-on software provides functions to convert facsimile files to graphics files or text files (and vice versa), software to convert facsimile into high-resolution graphics for laser printing, and terminal previewing. These packages require software to interpret inputs and fill in lost detail caused by the facsimile process.

Once in the PC or workstation, one can then access and use all the power of hosted tools to manipulate and edit the facsimile as would be capable with any stored compatible file.

Table 6-1 indicates some available LAN PC facsimile card vendors and the software features they support. The table is set up in paragraph form by company and will include the product name, printers that the product supports, standard file conversion, and software features.

LAN FILE SERVERS

In the previous section dealing with operating systems and distributed databases, we discussed some of the concepts needed for file servers. File server hardware is controlled typically by one of three file servers: Structures Network File Server by Sun Microsystems, Remote File Sharing by AT&T, or Server Message Block by IBM. These three software techniques for file servers are the main three used; they need to be examined by buyers, since the selection may impact future growth or limit additions.

Network File Server (NFS) implements a virtual file system by extending Unix's concept and adding the ability to redirect remotes, built-in external data representation, and remote procedure call techniques found in Unix network implementations to provide the lowest-level data handling.

The virtual file system of NFS takes operating system calls and checks each to see if it is a local or remote request. If local, the call is passed to the lower-level file server. If remote, the call is passed to lower-level NFS software that utilizes XDR and RPC protocols to implement access, synchronization, and transfers with remote NFS servers. The remote procedure call capability allows the originator to make a request to the server node to deliver a file to the caller. The XDR provides the mechanisms to compensate for data mismatches as to alignments, representations, etc.

AT&T's Remote File Sharing provides its features by setting up virtual streams and full duplex pipes (conduits) upon which data flows from client to server. As in NFS user service, calls from the operating system are provided to RFS. RFS sets up a virtual connection between the client and server providing a path for the control and data to flow. The RFS utilizes a name server to locate a server and to aid in the connection process. Once connected, the operating systems calls are directed to the remote file server process, which provides the actual transmission service. The key to this is the reliable virtual-circuit concept.

Table 6-1. Facsimile Products

Advanced Vision Research, Inc.
San Jose, CA (408)434-1115
Product: MegaFAX
Printers supported: Cannon, HP Laser
Standard files converted: ASCII,; PC Paintbrush; GEM; TIFF
Software features: Not applicable (N/A)

American Data Technology, Inc.
Pasadena, CA (818) 578-1339
Product: SmartFax
Printers supported: Epson FX, LQ; Cannon, HP Laser; IBM Graphics
Standard files converted: ASCII, TIFF, Epson FX 80, Dr Halo, PC Paintbrush, Wordstar
Software features: Help, Graphics Editor, Area Code Look-up, Blank-Page Creation

American Teleprocessing Corp.
Houston, TX (713) 973-1616
Product: Proto-Fax
Printers supported: Epson; HP Laser; QMS Laser; Okidata
Standard files converted: ASCII, N/A
Software features: Help, Editor, Forms Package

AT&T Information Systems
Morristown, NJ (201) 898-8000
Product: FAX Connection
Printers supported: Epson, HP Laser; AT&T 495; IBM Graphics
Standard files converted: ASCII Screen Capture
Software features: N/A

Brooktrout Technology, Inc.
Wellesley Hills, MA (617) 235-3026
Product: Fax-Mail 24
Fax-Mail 48
Fax-Mail 96
Printers supported: Epson; IBM; HP Laser
Standard files converted: ASCII, PC Paint; Fax to PC Paint
Software features: Text Editor, Graphics Editor

Brother International Corp.
Piscataway, NJ (201) 981-0300
Product: IntelliFAX
Printers supported: Brother Dot Matrix, Laser, Twinwriter 5 and 6
Standard files converted: ASCII, PC Paintbrush, Dr Halo, Most UP
Software features: Help, Graphics Editor, LAN Option

(Continued)

Table 6-1. *Continued*

The Complete PC, Inc.
Milpitas, CA (403) 434-0145
Product: The Complete FAX
Printers supported: Epson FX; HP Laser; Toshiba DOT; IBM Graphics; NEC
Standard files converted: ASCII, Epson FX-80, Dr Halo, PC Paintbrush, Microsoft Windows Paint
Software features: Help, Cover Page, Headers, Graphics Editor

Datacopy Corp
Mountain View, CA (405) 965-7900
Product: Microfax
Printers supported: Epson FX; HP Laser; Toshiba Dot; IBM Graphics, NEC
Standard files converted: ASCII, PC Paintbrush, TIFF, GEM
Software features: Help, Text Editor, Direct Print Incoming FAX

Datavue Corp.
Norcross, GA (404) 564-5668
Product: Datavue FAX Card
Printers supported: Epson Graftrax; HP Laser; Quadram Laser; Toshiba; Okidata
Standard files converted: ASCII (on the fly), Dr Halo, PC Paintbrush
Software features: Help, Graphics Editor

Dest Corp.
Milpitas, CA (408) 946-7100
Product: Facsimile Pac
Printers supported: Epson FX, HP Laserjet, Toshiba, Okidata
Standard files converted: ASCII, Epson FX-80
Software features: N/A

E.I.T Inc.
Fairfield, NJ (201) 22701447
Product: PC-FAX
Printers supported: Epson Dot Matrix; HP Laser
Standard files converted:ASCII, PC Paint
Software features: Help, Graphics Editor

Gammalink Corp.
Palo Alto, CA (415) 856-7421
Product: GammaFax, GammaFax MC
Printers supported: Epson; IBM; Cannon, HP, Xerox LASER; Fujitsu; NEC; Toshiba
Standard files converted: ASCII (128 or 256); Autocad PRP, Dr Halo, PC Paintbrush, TIFF
Software features: Help, Programmer's Toolbox, Windows, Background gateway for LANs

(Continued)

Table 6-1. *Continued*

GMS, A Division of Dest Corp.
Boca Raton, FL (305) 994-6500
Product: EZ-Fax
Printers supported: Epson FX, HP Laser; Toshiba P351; IBM Graphics; Okidata
Standard files converted: ASCII, Dr Halo, PC Paintbrush, Epson FX-80
Software features: Help, Intercepts Printer Stream, LAN Support

MicroTek Lab, Inc.
Gardena CA (213) 321-2121
Product: MFAX96P
Printers supported: Epson FX; HP, Canon, Cordata Laser
Standard files converted: ASCII (128 of 256), Wordstar, Eyestar Graphics
Software features: N/A

OAZ Communications, Inc.
Tustin, CA (714) 259-0909
Product: XaFax
Printers supported: Epson; HP, Canon Laser
Standard files converted: ASCII, TIFF, PC Paintbrush
Software features: Cover Page, Graphics Editor, Headers

Omnium Corp.
Stillwater, MN (612) 430-2060
Product: PC Fax System
Printers supported: Epson; IBM Graphics, HP Laser
Standard files converted: ASCII, PC Paintbrush, Dr Halo, Most WP
Software features: Help, Windowing, Text Editor

Panasonic Industrial Co.
Secaucus, NJ (201) 348-7000
Product: Fax Partner
Printer Supported: Epson; HP Laser; NEC; Toshiba; Panasonic
Standard files converted: ASCII, PC Paintbrush
Software features: Cover Page, Graphics Editor, Headers

Pitney Bowes, Inc.
Stamford CT (203) 351-6127
Product: Path II
Printers supported: HP, QMS Laser; Most Dot Matrix
Standard files converted: ASCII, PC Paintbrush
Software features: Graphics Editor, Windowing

(Continued)

Table 6-1. *Continued*

Printronix, Inc.
Dallas, TX (214) 630-9700
Product: DataFax
Printers supported: IBM; Canon, HP Laser; NEC
Standard files converted: ASCII, PC Paintbrush, Dr Halo, TIFF, Raster
Software features: VT-100 Emulation, Hot key to DOS, Text Editor

Quadram Corp.
Norcross, GA (404) 923-6666
Product: JT Fax 4800, JT Fax 9600
Printers supported: Epson; IBM; Canon, HP, Xerox Laser; Fujitsu; NEC; Toshiba
Standard files converted: ASCII, PC Paintbrush, Dr Halo
Software features: Help, Graphics Editor

Ricoh Corp.
West Caldwell, NJ (201) 882-2000
Product: ImageCard
Printers supported: HP, RICOH PC6000, Xerox 4045 Laser; Okidata
Standard files converted: ASCII, PC Paintbrush, Dr Halo, Graphic Screen, Saves
Software features: Store and Forward

Spectrafax Corp.
Naples, FL (813) 643-5060
Product: the FAXCARD
Printers supported: HP Laser; Postscript; Epson
Standard File Conversion: ASCII (on the fly), PC Paintbrush
Software features: LAN support, Graphics Editor

TEO Systems Inc.
Markham, Ont., Canada (416) 474-9372
Product: FCS Interface
Printers supported: Epson; NEC; HP Laser
Standard files converted: ASCII, PC Paintbrush
Software features: Help, Graphics Editor, Windowing

IBM's server message block protocol is DOS based and uses Microsoft's Redirector, the NetBIOS protocols, as its basic tools. Applications make calls to the local operating system. These calls are intercepted by the Redirector, which determines if this is a network access or local access. If it is a remote, the Redirector protocol uses NetBIOS software to set up peer communications. The NetBIOS software directs the call to the proper server which, using file server software, performs the wanted task.

Many vendors have utilized these basic capabilities to construct file servers. Interested readers are directed to the list of file server vendors in the appendix for more detailed information on their products built on these basic features.

NETWORK ADMINISTRATION SOFTWARE

Just as in the early days of computers and during their evolution, it has always been desirable to control and monitor systems usage. All have demonstrated the value of monitoring, auditing, and billback systems to control costs and bound requirements. Due to the relative infancy of most LAN environments, these features have not been built into the LAN operating systems environments. But vendors have been developing network monitoring and management packages for the LAN environment. Such systems provide services to monitor network traffic load, collect alarm conditions, collect diagnostic test results, network configuration, noise status, user statistics, line activities, status/resource maps, resource utilizations, and channel parameters, as some of their collected statistics. This data in turn is used to generate a variety of reports. Table 6-2 shows the various vendors surveyed and the type of management services offered by each. This is not a complete list nor does it list all of their capabilities.

Table 6-2 only outlines some basic capabilities of these tools for network management. Future network management systems will incorporate diagnostic tests and analysis, as well as more integrated monitoring of all components of a LAN or connected to the LAN. It can be seen that this problem of managing the network in itself is a distributed application and in time it will mature along with the LANs it services.

DIAGNOSTIC SOFTWARE

Related to management software, and in the future feeding it, is diagnostic software. Diagnostic tools in a LAN environment are used to locate faults in cables, hardware cards, and LAN software related to communications. Problems in corrections and cable shorts are some of the hardest problems to locate. One needs tools that allow the technician to "look" down the cable from end to end and see the input versus output. To do this requires both specialized hardware and software. Cable testers are used to isolate shorts and faults in cables. These devices have become so sophisticated that a trained technician and even LAN managers can use them to define the fault and where in the cable it resides. Another form of cable tester is the oscilloscope, which provides a means to test end-to-end performance by inserting signals and viewing their form on output. Deviations caused by noise sources and faults can be detected in this fashion.

To find the harder LAN faults requires special LAN testers called monitors. Monitors allow you to view data as they flow over the LAN to aid in diagnostic testing.

Table 6-2. Network Management Software

Vendor and Model: **AT&T Dataphone II**
Network Management Features:
Data recorded: Network status-device and facility faults, alarms, system access, test results, multiplexer traffic.
Reports generated: Network map, inventory profiles, trouble tickets, trending, user-defined color graphics, management reports.
Other network management services supported: Remote support all network devices/facilities by AT&T, automated trouble reports.

Vendor and Model: **Atlantic Research, NTS 3000 Network restoration, Test, and Management System**
Data recorded: User-definable alarms, trouble ticket, configuration, inventory, personnel databases, etc.
Reports generated: Unlimited user-defined reports on alarms, configurations, trouble tickets, inventories.
Other network management services supported: Other services supported via user-definable data management system.

Vendor and Model: **Avant-Garde Computing Net/Guard**
Data recorded: User activity alerts, session activation, port utilities audit trail, class profile.
Reports generated: Activity summary, current activity alerts, user profiles, port utilization.
Other Network management services supported: Access control, terminal emulation, protocol conversion.

Vendor and Model: **Blue Lance, Lt. Auditor**
Data recorded: All types of file I/O.
Reports generated: File and user I/O Report.

Vendor and Model: **Case Communications, Inc., Case 5200.**
Data recorded: Network configuration, hardware activity, line activity, alarms, histories, line troubles.
Reports generated: Network lists, trouble reports, activity summary, configuration ticket, trouble ticketing.
Other network management services supported: Diagnostics, soft strapping, central monitoring, dial backup, multiplexer support.

Vendor and Model: **Cincom Systems, Net/Master**
Data recorded: Configuration, activity, reference, inventory, hardware, alarms, model type alerts, operator, performance.
Reports generated: Network schematics, audit trails, trouble tickets, inventory, account summary, current activity, history, etc.
Other network management services supported: Failure, performance, configuration management; access control.

(Continued)

Table 6-2. *Continued*

Vendor and Model: **Codex Corp; 4840/4850/4860 Network Management Systems**
Data recorded: Configuration history, alarms, inventory, personnel, trouble tickets.
Reports generated: Alarms, history, inventory, user definable.
Other network management services supported: Monitoring, fallback, configuration.

Vendor and Model: **CRX Telcom Corp; System 100 Automatic Trunk Test System**
Data recorded: Trunk operative faults, transmission test results, out of specification flags for tests.
Reports generated: Trouble tickets, activity summary, demand test results, current activity.

Vendor and Model: **Computer Communications Co, LAN Map**
Data recorded: Configuration server status, user logs.
Reports generated: Status reports, trouble reports.

Vendor and Model: **Datacom, Management Sciences Anmacs**
Data recorded: Audit trail of all operator actions, alarms and abnormal conditions, and test result data.
Reports generated: Audit trail, as above; alarm status.
Other network management services supported: Automatic BERT and voltage/frequency level testing (Autoscan), remote control tests.

Vendor and Model: **Data Switch, Controlnet 200**
Data recorded: Configuration hardware, history, activity, security, and troubleshooting.
Reports generated: Trouble tickets, memo, network status, audit trail, configuration status.
Other network management services supported: Diagnostic, performance management, remote access, configuration management.

Vendor and Model: **Digital Communications Associates (DCA); NMS PC V2.0**
Data recorded: Configuration accounting, reference, inventory, hardware, alarms, activity history.
Reports generated: Network schematics, inventory, trouble ticket, alarms, audit trail, account summary, current account, history.
Other network management services supported: Failure, performance, configuration management, monitor, diagnostics.

Vendor and Model: **Dynatech, Data systems Dynanet 240**
Data recorded: Network configuration, hardware activity, line activity, alarms, histories, line troubles.

(Continued)

Table 6-2. *Continued*

Reports generated: Line, network, alarm activities, network configurations, trouble reports, diagnostic operations.
Other network management services supported: Remote switching, alarm monitoring, and test diagnostic access.

Vendor and Model: **EMCOM Corp., NCS70 Series**
Data recorded: Line utilization, line activity, response time, error and status real-time/retrieval, line trace/trap.
Reports generated: By line, controller device application, usage availability, current activity, message accounting.
Other network management services supported: Planning data, performance trends.

Vendor and Model: **FTP Software, LAN Watch**
Data recorded: Packet flow, content, source destination, length, errors.
Reports generated: Flow statistics, bottleneck statistics.

Vendor and Model: **General Datacomm Industries, NETCOM-7 NCM-70**
Data recorded: All network events, configuration, inventory.
Reports generated: Event summary, trouble tickets, inventory, configuration, graphics.
Other network management services supported: Host CPU interface, NMC-to-NMC interface, satellite controller.

Vendor and Model: **IBM, Netview**
Data recorded: Alarms, hardware.
Reports generated:
Other network management services supported: Hardware monitoring, session monitor, status monitor, control facility.

Vendor and Model: **Infinet, 90/70**
Data recorded: Configurations, hardware status activity, alarms, call record information, channel status, equipment ID.
Reports generated: Problem reports, trouble tickets, activity reports, inventory, accounting, network availability.
Other network management services supported: Performance management subsystem, automated escalation of alarms.

Vendor and Model: **Infotron Systems Corp., Advanced Network Manager (ANX INX)**
Data recorded: Configuration, events, status, operations message, call record information, alarms, channel status, equipment ID.
Reports generated: Configuration, alarm history, call record, channel status, current activity, trend profiles.
Other network management services supported: Diagnostics, primary console operation.

(Continued)

Table 6-2. *Continued*

Vendor and Model: **Infinet Performance Systems, Smart**
Data recorded: Configuration, response time utilities, availability, activity alarms, operations done, transaction type.
Reports generated: Alarms, activity, summaries current activity, history, exceptions available, throughput response time, transaction type.
Other network management services supported: Performance management, monitoring, configuration management.

Vendor and Model: **International Data Sciences (IDS) Series 9000 NM&TCS**
Data recorded: Configurations, activity, alarm, database.
Reports generated: Journal, activity, and statistical reports,
Other network management services supported: Distributed Intelligence.

Vendor and model: **Lanservices, Inc., Lantrail**
Data recorded: File activity, server maps, traffic flow.
Reports generated: Utilization, bottlenecks, statistical reports.

Vendor and model: **NEC America, Network Control and Management System (NCMS)**
Data recorded: Configuration, hardware, reference, alarms, inventory, trouble tickets.
Reports generated: Current and historical alarm, trouble ticket, modem attributes, modem, hardware, and circuit inventory.
Other network management services supported: Eye pattern diagnostics, BERT/BLERT, auto test, auto poll test, line parameters.

Vendor and model: **Northern Telecom/Spectron NMS**
Data recorded: Alarm, performance, data, audit trail, trouble tickets.
Reports generated: Exception reports, histograms on availability, performance, trouble tickets, and status.
Other network management services supported: Control of test equipment, T1 mux, and non-NT vendor network products.

Vendor and model: **Pacific Software, Network Assistant**
Data recorded: Printer queue, printer status, printer controls.
Report generated: Statistics

Vendor and Model: **Paradyne Analysis 6510**
Data recorded: Inventory, hardware, alarms, modem type, vendor, location, equipment.
Report generated: Trouble tickets, alarms, inventory, vendor performance, equipment performance, alert statistics, summary report.
Other network management services supported: Diagnostics.

(Continued)

Table 6-2. *Continued*

Vendor and model: **Racal-Milgo, CMS 2060 Series**

Data recorded: Configuration, activity, alarms, reference, inventory, hardware, modem type, operations performed.

Report generated: Network schematics, inventory, alarms, activity, summaries, current activity, history, performance trends.

Other network management services supported: Performance, monitor, diagnostics, configuration, line quality analysis.

Vendor and model: **Symplex, Maestro Network Management System**

Data recorded: Current status configuration.

Report generated: Activity summary, current activity, alarm conditions.

Other network management services supported: Failure and performance management.

Vendor and model: **Teleprocessing Products, Multidrop Network Manager**

Data recorded: Configuration, inventory failed scans, alarm archive, line files, drop files.

Report generated: Inventory, activity summary, current activity, line information, drop information.

Other network management services supported: Performance management.

Vendor and model: **3Com, Etherprobe**

Data recorded: Conversations, source/destination, packet transmission.

Report generated: Server loading, user traffic, network traffic.

Vendor and Model: **Timplex Link, Network Management Systems for Link/1 and Link/2 Systems**

Data recorded: Configuration, network port parameters and routing, operating statistics, and report generation.

Report generated: Network configuration, statistics, diagrams.

Other network management services supported: Downline or uploading of Link/1 or Link/2 system parameters.

Vendor and model: **Verilink Corporation, Verinet**

Data recorded: Configuration, activity, inventory, hardware, operations performed, alarms.

Reports generated: Network schematics, trouble ticket, inventory, history, alarms, audit trails, activity summary, current activity.

Other network management services supported: Failure, performance, configuration, management; monitor; diagnostics.

Vendor and model: **Westcon Assoc, ARC-Monitor +**

Data recorded: packets, configuration, node activity, tokens activity, flags.

Reports generated: Time graphs, configuration graphs, communications analysis-failure detection.

Table 6-3. LAN Diagnostic Tools

Manufacturer/Product	Function
Brightwork Development Arcmonitor	Tests interface card, coaxial cable and monitors traffic.
Vance Systems/ATS 1000	Protocol analyzer for token ring and bus networks.
Ontrack Computer Disk Manager/DOSUTIL	Disk Diagnostic software and optimizer.
Experdata/E20	Ethernet transceiver tester and driver.
Brightwork Development Emonitor	Ethernet-based LAN traffic monitor and analyzer.
LAN Tools/LANmap	Configuration control tool provides online mapping of configurations.
Legend software/LAN patrol	Traffic monitor and analyzer for Ethernet StarLAN networks.
EON systems/LAN probe	Provides a powerful tool to analyze Ethernet traffic and diagnose failures.
LAN Tools/LAN Traffic	A netware LAN traffic monitor.
Tektronix/OF150/235	Optical fiber testers for multinode and single node optical fiber links or LANs.
Network general/sniffer	Protocol analyzer for a wide range of LANs. Provides a wide array of tests and reports on LAN traffic, protocols.
Spider Systems/ Spider Monitor	Protocol analyzer for a wide array of LANs. Used to test protocol faults.
Proteon/Token View-4	Token ring network monitor and diagnostic tool.Aids in detecting hard and soft errors.
Experdata/VigiLAN	LAN performance monitor for Ethernet/StanLAN networks. Used for error detection.

These devices come in many forms, complexities, and price ranges based on their intended functions and diversity of applications.

Table 6-3 lists a sampling of some LAN diagnostic tools and their intended functions.

LAN DESIGN TOOLS

A local area network does not come into being by simply taking a piece of wire and plugging it from machine to machine. A LAN consists of transmission media, electronics to realize communications among connected machines, protocols that the hardware understands in order to converse intelligently, software to provide added features like logical naming, error detection and correction, link setup, peer synchronization, and a raft of other functions. With all these elements to select from, the LAN purchaser has a myriad of decisions and comparisons to make in the process of selection. These decisions must be made with the final user application base in mind. Once a LAN is selected, the adminstrator still must think about and be ready to adjust and grow the LAN based on new and constantly evolving requirements.

To perform these tradeoff analyses and be prepared to select new or added LAN components, administrators need data and analysis of the same. But, how do they get this information? Where does it come from? Administrators can perform the analysis themselves, but they require the tools to extract and construct data for the analysis. They could rely on vendors or third parties for the analysis, but this is not always the best choice. Whichever means is utilized, there is a requirement to have means to address network requirements. To be complete, this should follow traditional engineering practices such as structured design. If structured topdown design methodologies are utilized, then alternative architectures can be analyzed against each other over a spectrum of measures such as response time, cost, availability, reliability, extensibility, etc. The process takes into account present networking requirements and future growth projections.

The process involves three distinct phases: first is requirements definition which is followed by systems analysis and optimization, and finally by acquisition and lifecycle maintenance. Each phase has a distinct set of tools and techniques that can be applied. Requirements analysis typically involves collecting data on current communications, proposed growth, on hand equipments, future equipments, and intended uses. This data is then compiled and used to get a baseline of requirements that bound the range of performance and operational conditions that the future system must meet. Once a requirements specification has been developed, the administrator has defined what he or she thinks the LAN needs will be. The next phase is to collect and analyze alternatives.

During this phase some of the requirements defined during the previous phase may need another look and be adjusted. This occurs because tools applied to the analysis of the LAN provide feedback to LAN architects, who can then refine their assessments into viable designs. This process is aided through tools such as computer simulations, analytical models, and engineering tradeoff analysis. A detailed description of tools and techniques available to LAN designers is provided in the text *Modeling and Analysis of Local Area Networks* by Desrochers and Fortier, published in summer 1990 by Multiscience Press and CRC Press.

Network Design Tools

All the tools to be summarized provide either an analytical, simulation, or combined approach to modeling LAN alternatives. In all cases, the tools provide a means to input constraints and physical characteristics and then perform analysis based on various operational conditions. The output is an evaluation in terms of performance, cost, or some other measure of effectiveness that can then be used as a yardstick to measure the LAN alternatives gains.

The Aries Group provides a set of programs to perform network analysis using queuing theory, topological optimizations, data communications analysis, and tariff studies. An online database provides the needed data to perform the analysis.

The tools' emphasis is on aiding in the design of integrated voice and data communications networks and could be applied to LAN design. Output reports come in a variety from simple queuing to more elaborate tabulations and graphical formats.

AT&T offers a service to customers that provides leased-line communications network design aids. The tool, called E-INOS, provides a set of analytical modeling modules for communications designs. It provides for technology assessment as well as tariff assessment through extensive graphical interfaces.

BGS systems provides tools for designing (IBM) systems network architecture (SNA) based networks. These interactive tools can be used to aid in the description, evaluation, and performance prediction of components of an SNA network. Using these tools, designers can evaluate a variety of design alternatives quickly and efficiently.

Bridges and Assoc. markets a tools called HNDS (Hybrid Network Design System) that allows an analyst the capabilities to study real-world conditionals on network performance. Networks designed for are of the hybrid leased-line type.

Connections Telecommunications Products, Distributed Network Design System (DNDS), and Multipoint Network Design System (MNDS) were developed to offer aid in designing packet-switching irregular-topology networks. These tools aid network architects in the topological design and performance assessment design for such networks.

Like DNDS and MNDS, Contel Business Networks Modular Interactive Network Designer (MIND) provides capabilities to model and analyze distributed mesh networks based on leased-line type interconnects. The tool allows one to analyze topology design, capacity, tariffs, and protocols in developing a network.

CADNET is General Network Corporation's computer-aided design tool that provides a full interactive capability to develop, analyze, and optimize using simulations of large WANs and LANs. One can study configurations of equipment and topology, as well as transmission technology. This tool provides good graphical interfaces to make developing models and evaluating results fairly easy.

Quintessential Solutions developed a modular set of analysis tools and modules that provides designers with the means to customize specifications of the models to more realistically match projected or present systems conditions and to do the same with output modules. This is a detailed model providing for multilevels of analysis from high-level topological designs to packet/message communications analyst down to circuit design evaluation. This offers a wide range of design options to the analyst to model and evaluate in the selection process.

These tools, though, have not been specifically aimed at the LAN environment and therefore may not provide the best analysis of such environments.

A prototype tool described in *Modeling and Analysis of Local Area Networks* [Desrochers 1989] is a tool developed explicitly for the LAN environment. It provides a menu-driven interface to allow architects to construct a LAN topology, to populate it with devices, add to select parameters that define protocols, technology applied, and a myriad of other conditions. These can then be compiled into a simulation run that will evaluate the user's design. Loadings can be analytically derived or derived from actual measurements inputted by a scenario driver. It presently provides capabilities to model a large number of LAN protocols, topologies, media types, and attached devices. In addition, mechanisms exist for users to "add" their own models to the package to model yet unforeseen network types. Outputs are both tabular and graphical, providing the analyst with a variety of means to evaluate designs.

If this is still not flexible enough, the LAN developers can go to any number of generalized simulation languages to develop their own LAN simulations. Languages such as SLAMII, Network II.5, Simscript, and many more are available for use.

SECURITY

Related to both management and information components of a LAN is its security system. Security plays an important role in LANs since it, like any computer, control, manipulate, and disseminate an enterprise's most guarded secret—its information. The news has been full of problems and breaches of security within the LAN environments. How often have we heard of the university hacker breaking into some major company's databases via their LAN. Even the Defense Department has had its problems. In addition, beyond simple access and viewing, we have begun to see destructive access via infestation of LANs by computer viruses. These have brought down entire LANs and caused massive losses in productivity. To alleviate these and many other problems, LANs, as well as mainframes and the myriad other devices they serve, must be made secure. Products in support of LAN security are being marketed today, and more are coming into the marketplace everyday. Two such systems are Absolute Security Inc. products and Digital Pathways Inc. products. These two will be highlighted as examples. Interested readers are pointed to references in this text and to vendors for more information and further products.

Absolute Security Inc. markets a host of LAN security products that include LAN investigator, Lanstone, Lanaccess, and LANinvestigator-Plus. Lanaccess provides for control of access to the system and to its files and I/O subsystems. This component in conjunction with LANstone, which captures the keystroke activity of users by time, user, and date, providing for reproduction of activity for security audits and for detection of suspicious activity. LANinvestigator along with Investigator-Plus provide monitoring of program and file activity, and detection of changes to files and source code. This aids in the detection and isolation of viruses and deliberate sabotage, as well as a change-management audit trail.

Digital Pathways Inc. markets a wide range of LAN, WAN, and bridge security devices and software. Their products aim at controlling the access to devices versus monitoring and management after access is allowed. Their product protects against unauthorized access via direct connects, dial in, and bridge links. The packages come with support software to set up security boundaries, and configure the system, as well as tools to monitor and report activities of successful and unsuccessful users.

COMPUTER-ASSISTED DESIGN AND CONTROL MANAGEMENT SOFTWARE

To this point we have addressed some representative examples of LAN software for LAN management such as operating systems, management environments, data bases, security and information software such as file server software, office automation software, design software, and miscellaneous information distribution software. But as LANs progress, so does the software being ported to them and being developed for them. For example, LAN software has been developed for plant management and control, for group computer-aided design, for teleconferencing. All these became available due to the availability of LANs and support software on which to base them.

In this section we will summarize two plant management packages and two CAD packages.

Plant management for manufacturing facilities, distribution facilities, utilities, and many more have been needed for years. The problem in the past was that each distinct phase of operation such as materials, inventory, purchasing, production scheduling, personnel management, and many more, were all separate entities with their separate schemes for controlling and distributing information being applied. LANs provide a means to integrate these entities and further provide the framework to develop whole new products.

Navigator from Advantage Software provides a complete manufacturing management system. Developed around the Novell Network, this product provides functions and tools for all aspects of a plant's management.

It is a menu-driven system utilizing window displays to provide maximum access to information. The basic components are an integrating database that links tools for part identifications, product bill of material management, inventory management, purchasing management, materials planning, production planning, shop floor control, transaction history of activities, and report generation capabilities. The product provides the means for a manufacturing facility to link all aspects of its functions into a unified companywide management package.

Comac Systems Corporation markets a LAN-based tool called Comac for tracking and aiding maintenance management and other aspects of a company's operations. The product consists of tools for collection of data on capital assets, resource stock and purchases, personnel events, and an array of other plant control data. In addition, it provides a means to schedule and plan maintenance plans, emerging conditions, personnel schedules, parts, and availability. It also provides a means to track work in progress to monitor costs and quality of work. Using data collected by the products, plant histories and trends can be developed to assist in overall management and to keep the company flowing.

This is a very comprehensive plant management tool; it is installed and used over a wide user base. Details of this and other management products can be procured from Comac in Schaumberg, Illinois.

CAD/CAM

Computer-aided design and computer-aided manufacturing are becoming part of normal product developments. At first, stand-alone products came into being; now work-group products for LANs are coming. These provide team designs to be effectively managed and thereby improve design and manufacturing team productivity. Products that integrate the entire design process are here today, and better ones will be coming in the future. Two products reviewed include VG Systems Networking MicroCADAM and Autocads Networked Group version.

Vg Systems provides an environment under which organizations can share data without the need to copy or recreate the data. Using Vg CIM CAD, files can be shared and operated on by multiple CAD users simultaneously. CAD designs can be plotted and versions managed, all invisible to the average user. The integrated products allow designers, drafters, checkers, and publications to develop models, edit models, plot models, modify models, check models, transfer to mainframes, produce publications with integrated graphics and text, all within their LAN environments and with full access to all information. Vg Systems offers a full range of mainframe and microbased CAD design aids; these all can be linked over a LAN to provide a group design system. Details of this and other products can be obtained from Vg systems in Woodland Hills, California.

AutoCAD has offered some fine products for the microcomputer user for computer-aided design. It was natural as their uniprocessor product expanded that they would offer a network-based version of this highly successful tool. AutoCAD offers a menu-driven environment where users can select precarved shapes to build more complex shapes, or use freehand functions to construct a variety of designs. The networked version will provide all the functionality of the local uniprocessor version, but with the ability to share files and designs to allow work-group functionality. The new product should be quite a success, based on present AutoCAD installations.

APPLICATIONS DEVELOPMENT

The previous tools and products became available due to the ability of LANs to share data and coordinate activities. These basic software features are the ones that will provide for the upcoming explosion of LAN software applications products. In previous chapters, it was discussed how LAN operating systems and applications can be built on basic synchronization concepts such as message passing, object invocation,

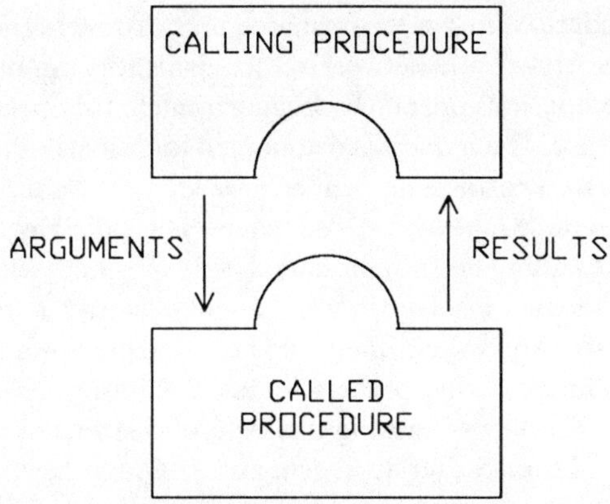

Figure 6-5. Uniprocessor Calling Mechanisms.

or remote procedure calling mechanisms. Although all these will work efficiently and provide the services needed for distributed applications, the more predominant one is the remote procedure call. Most of the major LAN vendors and operating systems vendors have advanced support of remote procedure calling capabilities. The basic reason vendors appear to be going this route is the simplicity of the mechanism and the ease with which it can be built and users can use it. Users write applications as procedures that call each other, pass parameters, and return results, just as in uniprocessor calling mechanisms (Figure 6-5). The difference is that the procedures must now interface to client/server processes and in networks to perform the task (Figure 6-6).

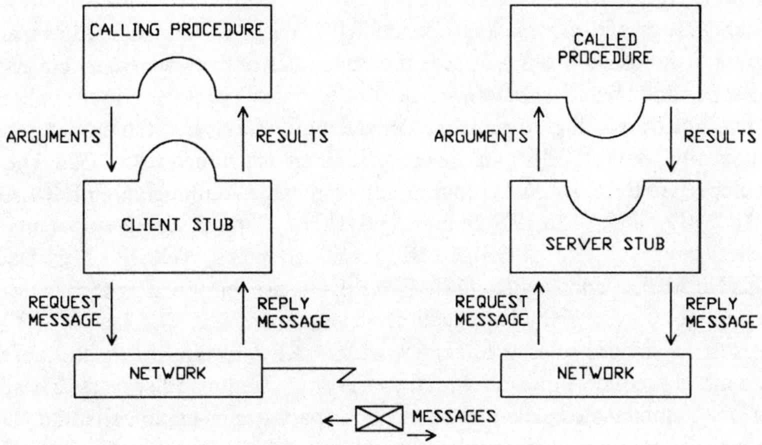

Figure 6-6. Remote Procedure Call.

The remote procedure calling mechanism requires extra servers in terms of clients for caller, servers for called, and network mechanisms for communications between them. Remote procedure calls operate by issuing a call to the operating system just as in the uniprocessor case. The request is determined to be local or remote. If remote, a client is invoked that locates the remote called procedure, and issues a request message to the server on the remote nodes host. The request is handled by the network, which transports it from the calling node to the called node (Figure 6-2).

The called node's network server issues the request message to the server on the called procedure node, who in turn relays this to the called procedure. One can see from this how simple the actual process is, but the implementation is something altogether different. The key to making this type of distributed processing concept work is to develop the client/server mechanisms. Once these are in place, user procedures simply issue and receive requests, perform their designed functions, and send and receive responses. Various networks and implementations use the basic concept of client and server in different ways.

In the DOS environment, the client and server need to have embedded in them the ability to call DOScallnmpipe which is a DOS LAN manager routine that opens a connection to the remote server, sends a request, waits for a reply, and then closes the connection. A logical name is required to point to the name of the server. On the server side, the DOS call DOSmakenmpipe opens the pipe for use. DOS connectpipe connects the server to the named client and DOS read causes a busy wait on the server waiting for incoming requests. On the client side, to issue a request for message the DOS write is used. DOSbufreset causes the client to block until the reply is received. To relinquish the RPC, the client issues DOSDissconnectnmpipe; in this fashion any set of cooperating clients and servers serving remote procedures and synchronization of them can be realized. Most tools are buys from vendors who, for remote procedure calls, provide compilers that perform the construction of the clients and servers for users, leaving users to be concerned with the application vice the network synchronization idiosyncracies.

Netwise Inc. provides a package called RPC Tools to help developers create network applications and keep them isolated from much of the worries associated with machine types and operating systems. The tool operates by taking a user application written in C and translating the network-dependent functions into the proper network application calls; normal calls in a user code segment are turned into RPCs. The tool was developed for the C and Ada programming languages running under DOS, OS/2, UNIX, ULTRIX, and VMS, VS Operating Systems. Supporting networks are NetBIOS, NetWare, Spx, Retix OSI/NetBIOS, named pipes, TCP/IP, TLI, Decnet, Banyans, Networks, and Touch OSI. Developers are provided with a powerful time-saving tool for distributed applications development.

Microsoft Corp. provides the framework for RPC by providing the network mechanisms via named pipes on which to develop clients and servers. These are the same mechanisms described in the previous paragraphs using calls like "DOS-calnmpipe" (servername/pipe/application name, RPC request buffer, and RPC response buffer).

Novell markets an RPC environment with a tool called RPC tool. Using this tool, distributed applications are built from applications code, network communications code, and network transport IPL protocol.

The Novell RPC tool provides developers with two options: the RPC tool can build the client and server code as well as the network code for the user, or the user can build their own RPC client and server code. As in Netwise's tool, this tool provides a powerful distributed applications support environment. It simplifies the development process, freeing up developers to concentrate on their applications development. The compiler for RPC tool automatically generates the source code for both client and server stubs. As in Netwise, users simply write their procedures and call routines allowing RPC tools compiler to generate the proper code.

The Netware RPC tool supports the C and Ada programming languages and runs under DOS and OS/2 environments.

COMMUNICATIONS SOFTWARE

LANs do not operate within a void. Most enterprises already possess some number of mainframes or minicomputers. It is natural to think that the PC LAN or miniLAN will be connected to this existing environment.

Software must exist to perform communications off-line by PCs to mainframes while users are gone, or to control sessions between active users on PCs and mainframes, or simply for communications only (message traffic, for example); still other equipment and software can provide LAN-to-LAN internetworking. The data transfer classes in these aforementioned interconnection means would be for file transfer only, electronic mail, direct active file access, or remote host (terminal line access).

The following section reviews some PC-based communications software to let PCs talk to mainframes.

Carbon Copy from Meridian Technology is a package that provides PCs phone linkup and access to files on a remote mainframe machine. In addition, it provides emulation capabilities to allow the PC to look and act like an attached terminal for remote access and operation.

Crosstalk from Digital Communications Associates is a high-powered PC-to-mainframe linkage package. This program has its own command language that provides for automatic machine-to-machine interaction. It has a text editor, many file transfer protocols, and supports a wide list of terminals in emulation mode. The features of this system offer it the capability to perform many jobs for LAN administrators, such as nightly backups of files and other processing at night while everyone is gone. Overall, this is a nice and very useful tool.

Hyperaccess from Hilgrave offers solid file transfer capabilities and supports Xmodemkermit terminal protocols. This is a simple package that operates well.

Instant Terminal markets a product that provides terminal emulation for a variety of vendor mainframes. It is a simple package and doesn't provide any extended services.

Maxonline is a menu-driven package that provides both file-transfer capabilities and terminal-emulation capabilities for a variety of vendors. This product provides a scripting language that allows users to write code to perform LAN to mainframe operations off-line.

Softklone markets Mirror II, which provides all the features of Crosstalk and adds a help system to aid learning, background mate, text editor, a wide variety of terminal emulations, and good screen displays. This is a powerful and versatile package and goes well beyond the basics of PC-to-mainframe linkage.

There are hundreds of other packages for terminal emulation and for PC-to-mainframe linkage. Most offer some number of emulations, provide a script language to allow users to program background access functions, and user aids such as learning tools, editors, and formatting programs.

More high-powered interconnects that provide LANs with many nodes access to other LANs, minis, and mainframes are called gateways or bridges. Products such as Novell's NetWare X.25 Point-to-Point Bridge, Multipoint Bridge and T-1 Bridges provide a variety of means and mechanisms to link LANs to WANs.

Banyan markets a network server that provides linkage of Banyan Vines Local Area Networks into mainframe environments.

A third gateway product is the Harris Network Communications System. This product, as the others, provides full linkage of LAN elements to mainframe elements. It provides full features such as terminal emulation, remote job entry, file transfer, and E-mail.

These products, as well as the myriad others not covered in this book, provide a means for PCs, minicomputers, and mainframes to coexist in an integrated fashion. Lists of vendors are supplied in the appendix and in previous chapters.

The final piece in the LAN puzzle is the integration of all equipment, even those from different vendors, into unified systems. To do this requires standard protocols for communications and standard means to format and view the translated information. The OSI (Open Systems Interconnection) protocols and standards have made this a possibility that will soon be a reality. This megaform of interconnection of everything to everything else has been coined as enterprise networking. The goal of enterprise networking is to provide a means of moving digitized information from any device to any other, regardless of manufacturer. This goal and the realization of this would open up the world to link all forms of business data online (Figure 6-7).

The push to realize true internetworking began with the development of the OSI protocols. These evolved and were refined by various working groups to develop the needed final standards. To move forward to true enterprise networking, a consortium of major vendors called the Corporation for Open Systems (COS), led by DEC, began and the MAP/TOP (manufacturing automation protocol/technical office protocol) user groups formed. These provide the focal point on which the standards are developed and promulgated. Of main interest to the group is the X.400 standard for message handling. This protocol is viewed as one of the first stops and a basic framework upon which the future interoperability standards, enterprise networking, and applications will be built. The players in this endeavor include Digital Equipment

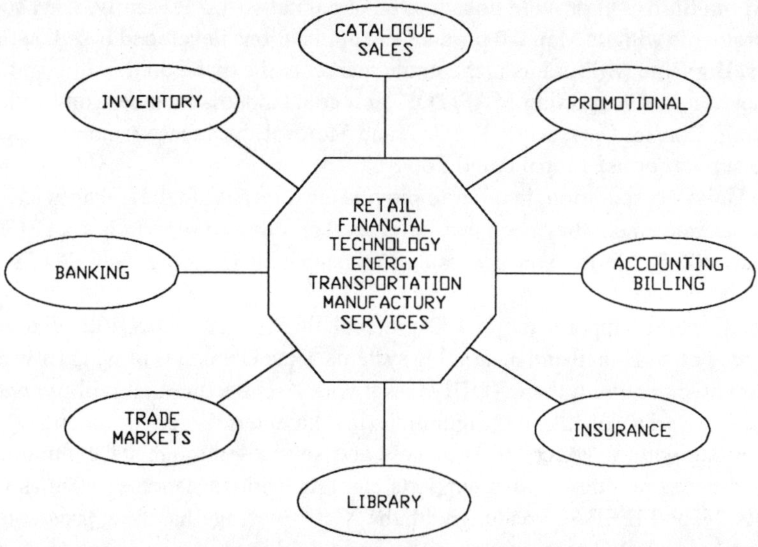

Figure 6-7. World-wide Enterprise Networking.

Corporation, Sun Microsystems, AT&T, Hewlett-Packard, Honeywell Bull, Telnet, IBM, Combustion Engineering, Control Data, Alan Bradley, Motorola, Unisys, Touch Communications, Moore Products Co., Data General, Mitsubishi, Wang, and many more.

Each brings to the endeavor the desire to see it come to reality and each is designing and manufacturing products to meet these goals.

DEC, for instance, provides products such as VAX OSI transport, mailbus, VAX OIS Applications Kernel, X.25 Gateway and message Router/X.400 deliver a framework upon which to produce future enterprise servers.

Sun Microsystems delivers OSI products such as Sunlink OSI, Sunlink X25, and Sunlink MHS that implement protocols to provide basic enterprise servers.

AT&T via its UNIX operating system has developed OSI products such as their Gateway 400 for X.400 protocol products, Accunet for X.25 packet service, and products such as E-mail and file servers that use these to communicate information to other X.400 or X.25-compatible vendor equipments.

Hewlett-Packard and Honeywell Bull have embraced the standards by implementing products to utilize the ISO 802.3 and CCITT X.25 standards. These products and others under development will allow these vendors to link to any member machines.

Telenet markets the Telmal 400 product. This product is based on the X.400 interconnect where Telenet was the principal architect. As before, this product will allow Telenet products to be linked and to converse with others who support X.400 interconnection.

IBM via SNA will provide links to OSI-based networks. Presently, IBM supports map protocols with its Map 3.0 product and applications developed based on this. As always, Big Blue will find its niche in the market and expand on it.

Other vendors supporting MAP/TOP protocol standards include Combustion Engineering, Control Data, Allen Bradley, and Motorola. All these vendors have products to support or use Map 3.0 and Top 3.0.

The Unisys corporation, in order to provide customers with the linkages and future product enhancement they need, has developed products to support X.25, IEEE LAN Standard 802.3, file transfer access and management (FTAM), and CCITT X.400 standards.

Data General supports the X.400 protocol through its DG/X.400 protocol and products such as E-mail and messaging systems. Data General is moving forward into LAN involvement with their XODIAC Network Architecture that will provide integration of a variety of LAN configurations into an enterprise environment.

Wang supports a variety of protocols and services through its commitment to standard-based products. Wang supports communications standards such as CCITT X.21, X.25, and IEEE 802.3 along with the X.400 message-handling protocol.

Wang has invested heavily in developing networking products based around these standards to keep a competitive edge.

Companies that will survive and grow into the coming century must let go of their closed-proprietary protocols and networks, and migrate or evolve into the standard protocols and networks. The ones that do will prosper; others will fall off to the side as industry, governments, and educators migrate toward the integration of all computing and informational assets into large interconnected computational environments.

SUMMARY

This chapter provided an introduction to the variety of software presently available for the LAN environment. Covered were software products for managing the LAN, providing basic communications, providing operating systems services, database services, file access, mail service, office automation, specialized office aids, CAD/CAM simulation, design aids, languages, and distributed applications development protocols. In all, the LAN environment is expanding and the software to meet the growing demand is coming to market. The future will see great strides in new software that will provide worldwide information and processing assets to anyone connected to a LAN. The challenge is to select applications areas and develop these new and exciting products.

This book provided a cursory look at the breadth of today's LAN software marketplace. There are literally thousands of vendors who market LAN products, although up to a few years ago most were simply wrapped around the communications and basic services areas. The trend now is to use these services to build innovative products that take advantage of the distribution to provide more effective tools for business, manufacturing, control, government, and educational endeavors.

APPENDIX

CATEGORIZATION OF LAN SOFTWARE PRODUCTS

INTRODUCTION

Local area networks and the software that supports them exist as a means to support end-user objectives of information transfer and management. The LAN and its software provide services to exchange digitized text, data, graphics, images, and voice, as well as many other forms of information. Additionally, the LAN provides the ability for applications to control the execution of other applications on many different sites. That is, the LAN software provides the processes to allow the combining of computer and communications equipment from a variety of vendors into a unified, usable environment.

The "information network" is the realization of integrated hardware and software from multiple vendors into one working system. Information from one machine is readily accessible as it is processed on the various machines. An information network provides access to all aspects of an enterprise's information bank (Figure A-1). Such an integrated system with all the classes of software available would provide an enterprise with formidable capability to compete in today's information-dependent market. With such a network, salesmen or field representatives can send along customer product suggestions or improvements directly to the design team or research and development teams. Information for decisions can be collected and collated from all sources quickly, providing potentially the edge one needs to win in today's markets. Such an integrated environment would allow an enterprise to work more efficiently and with the most up-to-date information. There are no more delays in conventional mailing of information, all transactions can be done online and all information shared, keeping the entire enterprise current.

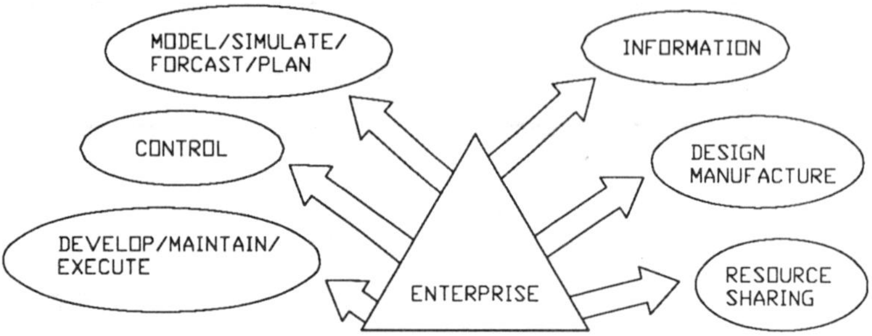

Figure A-1. Information Network Functional Categories.

Beyond the company, networking with customers, suppliers, etc., can provide even a better edge. Before we realize a totally integrated environment, the basic features need developing. Many strides have been made to this end, with new products coming onto the scene daily. In the not-so-distant future, high-powered environments will be here. The following sections will introduce the various catalogs of LAN software and provide an outline of some available products. The discussion will be kept at a high level in this chapter. Details of some products were presented in Chapter 6.

INFORMATION

This category of LAN software is probably the one we all know the most about. This is due to it being one of the first exhibiting products. In particular, we have all at some point used or heard of electronic mail. Electronic mail, though, is only a small part of the products found in this category. Additional products such as graphics tools, bulletin boards, report generators, teleconferencing facilities, database servers, and facsimile procedures make up some of the others.

Since mail services were one of the first, we will address that category first.

Electronic Mail

Electronic mail is a simple application that allows users to send any type of data from one site to another site. Control of the transfer is handled in a straightforward fashion. The sender provides a destination address and the message to be sent (either straight text or a data file). The address is packaged with the message into a mail unit (Figure A-2). The mail unit is sent through the system like a datagram, leaving error detection,

Figure A-2. Mail Unit.

etc., to the users. This mechanism provides a means to simply send raw data or anything else with few controls and checks. The mail system takes mail from the local mail repository and sends it to the destination repository. Users who wish to view the mail must enter the mail subsystem and use its features to read mail or, if wanted, delete mail, edit mail, create mail, send mail, file mail, sort mail, or many other features that may be provided.

The variety of mail services and what they provide is as diverse as the set of vendors shown in Table A-1 that provide such services. Table A-1 contains vendors who indicate that they have mail services, although there was no attempt to extract details. Not all vendors are included, and I apologize for omissions. If your favorite is not here, please drop me a line so that I can incorporate them in future editions. A few mail systems were reviewed in more detail in Chapter 6.

File/Text Transfer

File and text transfers are services similar to those of an electronic mailer in that they, too, must send information from one user to another. The difference is in how the services are performed, and where the end transfer goes. Remember, in the electronic mail systems mail is sent from user mailer to recipient mailbox. There is no notion of a usable file—just a piece of mail that a user can pick up when he or she is ready. A file server, on the other hand, transfers entire files from one location to another. The files can be copied or simply moved according to the user's discretion. Additionally, most file servers include mechanisms to guarantee correct transfer. This implies built-in use of error detection/correction or recovery techniques.

File transfer software can run the gamut from simple capability where the users must do all the legwork (i.e., set up route, know destination site and filename, and how to operate through the network) to a distributed file system that handles the transfer in a transparent fashion among the logical users. In any case, this service is one of the earlier ones to arrive onto the local area networking scene, and numerous vendors provide such packages. Table A-2 lists the vendors that provide file transfer or text transfer packages for either LAN-based or TI-based networks. Details of some of these file transfer packages were provided in Chapter 6 as indicated previously. I apologize if I have left out any vendors and would appreciate receiving additions to this list so that it may be as consistent as possible.

Table A-1. Electronic Mail Vendors*

ADAX, Inc.
ADR, Inc.
AIDCOM Assoc.
American International Communications Corp.
American Teleprocessing Corp.
Amtel Systems Corp.
Apple Computer Inc.
Applied Communications, Inc.
Applix, Inc.
Apollo, Inc.
ASC Computer Systems
AT&T Business Markets Group
Autotel Information Systems
BTI Computer Systems
CCI
Cincom Systems Inc.
Compuserve, Inc.
Computer Projects Inc.
Comterm, Inc.
Comverse Tech Inc.
Connections Telecommunications Inc.
Consumers Software Inc.
Convergent Technologies
Cross Information Co.
Cyber Digital Inc.
Data Architects Inc.
Data Rentals/Sales, Inc.
Data Com Solution, Inc.
Data Point Corp.
Dialcom Inc.
Digital Equipment Corp.
Eicon Technology Corp.
Electronic Interface Assoc. Inc.
Electronic Mail Corp of America
Emerald City Inc.
Ericsson Information Systems AB
Extel Corp.
Flexlink International Corp.
Frontier Technologies
FTC Communications, Inc.
Gandalf Data Inc.
Gateway Communications Inc.
Graphnet Inc.
Lanier Business Products/Sector of Harris Corp.
Harris Corp.
Hawkeye Grafix Inc.
Hewlett-Packard Co. Business Systems Sector
Honeywell Bull, Inc.
IBM
Incomnet Inc.
Infonet Computer Services
Intel Corp.
Intelligent Business Comm.
International Research and Evaluation
Lanquest Group
Lex Computer Systems
Linkware Corp.
Mai Basic Four Inc.
Marconi Instruments Ltd.
McDonnell Douglas/Mctel Inc.
Micro Systems Software
Miracle Technology UK Ltd.
Modern Technology Inc.
Motorola Inc.
Multicomm Telecommunications Corp.
Network Solutions
Nokia Information Systems
Northern Telecom Inc.
Novell
Omnicom Inc.
Online Software Int'l. Inc.
Orion Network Systems, Inc.
OST, Inc.
OTC Australia
PC Communications Ltd.
Philips International
Point-4 Data Corp.
QL Systems Limited
Quantum Software Systems Ltd.
RCA Inc.
Retix
Rezi Internationales
Ricoh Corp.
S&H Computer Systems Inc.
Satellite Technology Management Inc.

(Continued)

Table A-1. *Continued*

SCICOM Research US	Texas Instruments, Inc.
Softswitch Inc.	Titn Inc.
Soft Ronics	Touchbase Systems Inc.
SST Data Inc.	Transend Corp.
Sun Microsystems Inc.	Travtech Inc.
Sydney Development Corp.	Unisys
Sydney Ltd.	US Sprint
Syntax	Verimation Inc.
Syntrex Inc.	Videodial Inc.
Syscom Inc.	Wang Laboratories, Inc.
System Strategies Inc.	Waterloo Microsystems, Inc.
Sytek Inc.	Wells American Corp.
Tandata PLC	West Electronics Ltd.
Tandy Corp./Radio Shack	Western Digital Corp.
Technological Software Concepts	Western Union Telegraph Co.
Telamon Inc.	Wil Tek Inc.
Telenet Communications	Winter Halter Inc.
Telindus NV	

*Table A-1 lists the vendors that provide message-based mail services over LANs and TI links.

Table A-2. File/Text Transfer Package Vendors

ADAX Inc.	Cincom Systems Inc.
Agile	Communications Solution Inc.
ALTX Inc.	Comterm Inc.
American International Communications Corp.	Concept Development Systems Inc.
	Consumers Software Inc.
Apple Computer Inc.	Continental Resources Inc.
ARC Data Systems Inc.	Corporate Microsystems Inc.
Architectural Communications Inc.	CPT Corp
ASC Computer Systems	Cross Talk Communications/DCA
Asher Technologies	Data Interface Systems
AT&T	Data Rentals/Sales Inc.
Attachmate Corp.	DataCom Solution Inc.
AVNET Computer Technologies Inc.	Datanex Inc.
Banyan Systems Inc.	Data Point Corp.
Bell Communications Research	Dayna Communications
Bizcomp Corp.	Digital Equipment Corp.
Bridge Communications	Dove Computer Corp.
Browns Operating Systems Services Ltd.	DTSS Inc.

(Continued)

Table A-2. *Continued*

Eastman Communications	Persoft Inc.
Excelan Inc.	Point-4 Data Corp.
Flagstaff Engineering Inc.	Point of Sale Systems
Flexlink International Corp.	Polygon Inc.
Frontier Technologies	Prime Computer
Gateway Communications Inc.	The Protocol Team Inc.
Gateway Microsystems Inc.	Pyramid Technology Corp.
General Micro Systems	RCA
Harris Corp.	Retix
Hawkeye Grafix Inc.	Rose Electronics
Hewlett Packard Co.	S&H Computer Systems Inc.
Honeywell Bull Inc.	Softronics
IBM	SST Data Inc.
Intel Corp Systems Group	Sterling Software
Interlink Computer Sciences	Sun Microsystems Inc.
IO Corp.	Sydney Development Corp.
Laguna Labs	Syntax
Lex Computer Systems	Syscom Inc.
Linkware Corp	System Strategies Inc.
MAI Basic Four Inc.	The Systems Center Inc.
Marconi Instruments Ltd.	Talon Technology
Martin Marietta Data Systems	Tandy Corp
McDonnell Douglas	Techland Systems Blue Lynx
Meridian Technology Inc.	Telamon Inc.
MiCom Inter LAN	Telcor Systems Corp.
Micro Systems Software	Telexpress Inc.
Miracle Technology UK Ltd.	Titn Inc.
Motorola Inc.	Torus Systems
Mux Lab Inc.	Touch Communications Inc.
Nestar Systems Inc.	TRAX Softworks Inc.
Network Software Assoc.	TSI
Network Systems Corp.	Ultranet Ltd.
Nokia Information Systems	Ungermann-Bass Inc.
Northern Telecomm Inc.	Unique Automation Prod.
Norton Lambert Corp.	Unisys
Novell Inc.	Vector International\Virtual Microsystems
Omnicom Inc.	VM Personal Computing Inc.
On-line Software Inc.	Walker Richer and Quinn Inc.
On-line Computer Business Systems	Wang Laboratories Inc.
PC Communications Ltd.	Western Digital Corp
The Pande Group Inc.	Winterhalter Inc.
Pathway Design	XIOS Systems Corp.

PC-TO-MAINFRAME COMMUNICATIONS SERVERS

Communications software in early systems was limited to dial-in lines and terminal-emulation packages. Today's packages provide communications for a wide range of processor types and interfaces. There are today communications software packages to allow a variety of PCs to talk to each other, as well as to talk to minicomputers or mainframes. Likewise, there is a wide range of services built on top of this basic service. Typically, these packages utilize existing I/O points with software to emulate the device that would typically be attached to a terminal or a NIU. Table A-3 lists the vendors who indicated that they have products that allow PCs to talk to mainframes or vice versa. Details of a few products are in Chapter 6.

WORD-PROCESSING SOFTWARE

Early word processors did little more than allow the user to type documents and adjust text to keep margins accurate. As time went on they acquired much more power and diversity of functions, allowing for production of elaborate documents with illustrations, formatting, special characters, etc. But basically they still provide a simple service, that of document preparation assistance. In such capacity, they provide capabilities to enter text and edit text, to mark and move text, to find and replace specific text, and to format or reformat the presentation of the documents. They allow special commands to control the printing process and aid in the formatting and construction of tables. Additionally, they provide for self-control of files to save files and backup files for recovery. All of the packages offer some form of these basic features while some offer elaborate additions for figures, icons, etc. These more advanced packages will be referred to as publishing software, which is included in this section also. Table A-4 contains vendors who provide basic word-processing packages, and Table A-5 provides a list of vendors who provide publishing packages.

GRAPHICS PACKAGES

Software for producing graphics has been growing at a rapid pace. Additionally, these packages have begun to utilize more standards, with the result being that graphics can now be shared among many users and transmitted over the network. This also implies the ability possibly to utilize remote graphics servers to provide for use by more than one user. Graphics packages take on different forms; for example, some can be strictly freeform drawing such as Paint or Draw. Others offer the use of freeform mixed with

Table A-3. PC Communications Software

ACCU Sort Systems Inc.
ADR Inc.
Agile
AIDCOM Associates
ALTOS Computer Systems
American International Communications Corp.
American Teleprocessing
ARC Data Systems Inc.
Architectural Communications Inc.
Astrocom Corp.
AT&T
Attachmate Corp.
AVNET Computer Technologies Inc.
Banyan Systems Inc.
Barr Systems Inc
Bell Communications Research
Bizcomp Corp.
Bridge Communications
Cambridge Computer Associates Inc.
Carleton Corp.
Channel Systems International Inc.
Cleo Software
COMCAB
Communications Sciences
Communications Research Group
Communications Solution Inc.
Computer Projects Inc.
Comterm Inc.
Concept Development Systems Inc.
Consumers Software Inc.
Continental Resources Inc.
Corporate Microsystems Inc.
CPT Corp.
Cross Information Co.
Crosstalk Communications/DCA
Data/Com South Inc
Data Rentals/Sales Inc.
Data Com Solution Inc.
Data Point Corp.
DAVOX Corp.
DAYNA Communications Decision Data Computer Corp.
Digicom SPA
Digital Equipment Corp.
Dow Jones News/Retrieval
DTSS Inc.
EI Con Technology Corp.
Ericsson Information Systems AB
ESCA Inc.
Fox Research
Frontier Technologies
FTC Communications Inc.
FTP Software Inc.
Gateway Microsystems Inc.
General Digital Corp.
General Micro Systems
Lanier Business Products
Harris Corp.
Hawkeye Grafix Inc.
Hewlett-Packard Co.
Honeywell Bull Inc.
IBM
Information Builders Inc.
Intel Corp.
Intercomputer Communications Corp.
Jeumont Schneider
JMI Software Consultants
Laguna Labs
Lex Computer Systems
Linkware Corp.
MAI Basic Four Inc.
Marconi Instruments Ltd.
Martin Marietta Data Systems
MBS Communications Ltd.
McDonnell Douglas
Meridian Technology Corp.
Micro-Inter LAN
Micro Plus
Micro Systems Software
Micro Tempus Inc.
MicroStar Software Ltd.
Miracle Technology UK Ltd.
MUX Lab Inc.
Network Software Assoc.
Nokia Information Systems
Northern Telecom Inc.
Norton-Lambert Corp.
Novell Inc.
Office Solutions Co.

(Continued)

Table A-3. *Continued*

Omnicom Inc.	Talon Technology
Omzig Corp.	Tandy
On-Line Software International Inc.	Techland Systems
Online Computer Business Systems	Tele-gem
OST Inc.	Telexpress Inc.
PC Communications Ltd.	Titn Inc.
Pacer Software Inc.	Touch Communications
Packet/PC Inc.	Transend Corp
Pathway Design	Travtech Inc.
Persoft Inc.	Ultranet Ltd.
Point of Sale Systems	Ungermann-Bass Inc.
Polygon Inc.	Unique Automation products
RCA Inc.	Unisys
Retix	Urgeo Software Inc.
Rose Electronics	Virtual Microsystems
Softronics	VM Personal computing Inc.
Software AG of North America	Walker Richer and Quinn Inc.
Sterling Software	Wang Laboratories Inc.
Sun Microsystems	Waterloo Microsystems Inc.
Syntax	West Electronics Ltd.
Syscom Inc.	WinterHalter Inc.
The Systems Center Inc.	Xerox Corp.
SyTek Inc.	XIOS Systems Corp.

icons that can be selected, reproduced, scaled, rotated, etc. The results are that there exists a wide range of packages for graphics production. The users must select the one that best suits their needs. Chapter 6 goes into detail on one such package, whereas this appendix will but list the vendors who indicate that they have such tools. Table A-6 lists those vendors.

FACSIMILE

Facsimile machines have been with us for quite some time, and have proven their worth as a quick and cheap way to send hard copy from one place to another. Facsimile is growing into a new world with the advent of the computer and networking. It has become part of an integrated business information environment.

Table A-4. Word-Processing Software Vendors

Advanced Computer Technology	LEX Computer Systems
Apple Computer, Inc.	MAI Basic Four Inc.
Applix Inc.	Martin Marietta Data Systems
Ashton-Tate	MicroPro Inc.
AT&T	Microsoft Inc.
AVNet Computer Technologies Inc.	MicroSystems Software
Bell and Howell	Micro Data Base Systems Inc.
Boeing Computer Services Co.	Nokia Information Systems
Computer Consoles Inc.	Noumenon Corp.
ComTerm Inc.	Point-4 Data Corp.
Contel	Popular Programs Inc.
Continental Resources Inc.	Quebec Telephone
Cyma/McGraw Hill	RCA
Data Processing Design Inc.	Recognition Equipment Inc.
Data Rentals/Sales Inc.	The Santa Cruz Operation Inc.
Data Point Corp.	Solution Ware
Decision Data Computer	Southwestern Bell Telecom
Digital Equipment Corp.	Sun Microsystems Inc.
Electronic Interface Association Inc.	Supersoft Inc.
Emerald City Inc.	Sydney Development Corp.
Ericsson Information Systems AB	Syntrex Inc.
GTE Data Services Inc.	Tandy Corp.
GTE Supply	Texas Instruments Data Systems Group
Lanier Business Products	Trax Softworks Inc.
Harris Corp.	Unisys
Honeywell Bull Inc.	Vendurcom Inc.
IBM	Wang Laboratories Inc.
Information Processing Techniques Corp.	Xtend Communications
Inforonics Inc.	Zenith Data Systems
Innovative Software Inc.	Zylab Corp.
JMI Software Consultants Inc.	

The facsimile market represents approximately 1.5 billion dollars of sales per year and as such is a lucrative market to be in. More important, with PCs coming of age, integrating facsimile services into these and providing this service over one's network could be a real profit booster.

Facsimile software, along with a basic hardware element, provides for file management, storage of telephone numbers, and translation services to format nonfacsimile data into facsimile.

Table A-5. Publishing Software Vendors

Kurta Corp.
Motorola Information Systems
Science Accessories Corp.
Apple Computer Inc.
Adobe Systems Inc.

The following Table A-7 lists the vendors who supply facsimile equipment and software for use in a WAN/LAN environment.

DATABASE MANAGEMENT SYSTEMS

An important, if not the most important, component of an enterprise's informational assets is its database management systems. Database management systems provide services to manage data reads and writes in such a way as to guarantee the correctness of the information over time. The data manager protects consistency by checking the boundaries and type of updates being attempted, and provides various forms of user interface. Databases on local area networks provide another level of service. DBMS's provide a corporation's entire database assets to all; they provide management with tools to track and control how data is used, while increasing overall performance, security, and reliability of the corporate information assets. Table A-8 lists vendors who indicate that they either have network database servers (i.e., nodes that can be accessed remotely), or distributed database capabilities. An example of each style was given in more detail in Chapter 6.

Table A-6. Graphics Packages Vendors

Apple	Microsoft Corp.
Digital Equipment Corp.	Omnifax/teleautograph
Graphon Corp.	Panafax Corp.
Harris Corp.	Science Accessories Corp.
IBM	Sun Microsystems
Intergraph Corp.	Unisys
Microvitec Inc.	Wang Laboratories, Inc.

Table A-7. Facsimile Vendors

Advanced Vision Research Inc.	Omnifax/Teleautograph
American Data Technology Inc.	Omnium Corp.
American Teleprocessing Corp.	Online Computer Business Systems
ARDY Produkter AB Ardy Electronik	OTC Australia
ASC Computer Systems	Panafax Corp.
AT&T	Panasonic Industrial Co.
Brooktrout Technology Inc.	Pitney Bowes Inc.
Brother International Corp.	Pritronix Inc.
The Complete PC Inc.	Quadram Corp.
Data Copy Corp.	RCA Inc.
Data Vue Corp.	Ricoh Corp.
Dest Corp.	Sanyo Business Systems Corp.
EIT Inc.	Satellite Technology Management Inc.
Electronic Mail Corp of America	Southwestern Bell Telecomm
Extel Corp.	Spectrafax Corp.
Flagstaff Engineering Inc.	Sydney Development Corp.
Gammalink Corp.	Telecomet
GE American Communications Inc.	TEO Systems Inc.
GMS, A Division of Dest Corp.	3M Faxxchange Dept Office Systems Div.
Lane Telecommunications Inc.	Titn Inc.
Microtek Lab Inc.	Toyomenka Inc
Murata Business Systems	Xerox Corp.
OAZ Communications Inc.	

VIDEO TELECONFERENCING

The last but one of the most recent advances in online information services is video teleconferencing. This form of LAN information exchange takes on various forms. The service provides for voice, data, and visual transfer of information over dedicated lines where users can hold meetings remotely from each other over the electronic media. Table A-9 lists vendors who supply video teleconferencing services, equipment, and software.

RESOURCE SHARING

Resource sharing, as we saw in previous chapters, requires low-level systems operators to work, such as remote procedure calls, remote invocation, message passing, and the protocols associated with how to use these methods to provide transport resource sharing among users. This section will deal with software that provides these basic

Table A-8. Database Servers/Distributed Database Vendors

Amperif Corp.	International Research and Evaluation
Applications Software inc.	LAN Quest Group
Applied Data Research Inc.	Magnetic Press Inc.
Boeing Computer Services Inc.	MAI Basic Four Inc.
Bradmark Computer	Martin Marietta Data Systems
Britton-Lee Inc.	Micro Database Systems Inc.
BRS Information Technologies	Microforms Trans-lingual
CIMpoint	Must Software International
Complete Computer Systems	NADEK Computer Systems Inc.
CRI Inc.	Nantucket, Inc.
Campus America Inc.	National Data Corp.
Century Analysis Inc.	National Information Systems Inc.
Cincom Systems Inc.	Nestar Systems Inc.
Cognos	Newsnet Inc.
Compuserve Data	NJK Associates
Computer Associates International Inc.	Northern Telecom Inc.
Computer Corporation of America	Office Smiths Inc.
Concurrent Computer Corp.	Online Software International Inc.
Cullinet Software, Inc.	Oracle Corp.
Data Com Solution Inc.	Orion Network Systems Inc.
Data General Corp.	PAIS Inc.
Digital Equipment Corp.	Prime Computer Inc.
DBlaccess Inc.	Public Office Corp.
Dow Jones News/Retrieval	Quodata Corp.
Exact Systems and Programming Corp.	RR Bowker
Financial Technologies	RCA Inc.
Fulcrum Technologies Inc.	Relational Technology Inc.
General Data Systems Inc.	Rhodnius Inc.
Gentry Inc.	Ruf Corp.
Gupta Technologies	SAS Institute Inc.
Harris Corp.	Saturn Systems Inc.
Henco Software Inc.	Seed Software Corp.
Hewlett-Packard Co.	Signal Technology Inc.
Honeywell Bull Inc.	Shirley Institute
IBM	Software A.6 of North America Inc.
Info Globe	Sun Microsystems Inc.
Information Builders Inc.	Sybase Inc.
Information Dimensions Inc.	Tandem Computers Inc.
Information Structures Inc.	Tandy Corp.
Informix Software Inc.	Telegem
Inforonics Inc.	The Ultimate Corp.
Intel Corp Systems Group	Texas Instruments Data Systems Group
Interbase	Unify Corp
International Parallel	Unisys Inc.

Table A-9. Video Teleconferencing Vendors

American Laser Systems Inc.	OST Inc.
Brown Operating System Services Ltd.	OTC Australia
Cambridge Computer Associates Inc.	PC Communications Ltd.
Couid Inc.	RCA Global Communications Inc.
Data Cube Inc.	Shore Brothers Inc.
Data Point Corp.	Tandata PLC
Electrohome Ltd.	Tandy Corp.
FTC Communications Inc.	Telecomet
GE American Communications Inc.	TSI
Interand Corp.	Unisys Corp.
International Research and Evaluation	US Sprint
Microstar Software Ltd.	VideoDial Inc.
Nokia Information Systems	Wang Laboratories Inc.
Optel Communications Inc.	

services, as well as operating systems, distributed data-processing applications software, and simple remote server stations.

Operating Systems

Operating systems for LAN or WAN use are of two species. One is the network operating system variety, where the network portions are addressed as an afterthought or add-on and the users are required to know something about what they are doing. Or it could be of the distributed operating system variety where the operating system is fully integrated over all stations and the user has a uniprocessor (albeit a huge one) view of the system. In either case, the operating systems have the job of managing systems assets and providing users with their interface into the system. Typical functions include memory management, I/O management, server management, remote access, fault tolerance, network administration, security, and other such service functions.

Table A-10 lists vendors who provide operating systems or pieces thereof that provide some level of access to LAN/WAN services. Details of a few operating systems were addressed in Chapter 6.

Distributed Data-Processing Software

Software in this category is typically a cross between operating system and specific server types. The function of software within this category is to provide the basic

Table A-10. Network/LAN Operating Systems Vendors

Apollo Computer
Apple Computer Inc.
Artisoft Inc.
AT&T
Banyan
BBN Communications Corp.
Bell Communications Research
CBIS Inc.
CMC
COMCAB
COMPULAB Corp. Sentinel Div.
Computoll Group Ltd.
Corvus Systems Inc.
Continental Resources Inc.
Convergent Technologies
Data Rentals/Sales, Inc.
Data Point Corp.
Dataviz Inc.
Digital Equipment Corp.
EasyNet Systems Inc.
Ericsson Information Systems AB
Excelan Inc.
Frontier Technologies
FTP Software Inc.
Goal Systems International
Kinetics Inc.
Harris Corp.
Hewlett-Packard Co.
Honeywell Bull Inc.
IBM
Information Processing Techniques Corp.
Intel Corp.
JMI Software Consultants Inc.
Lanier Business Products
Lotus Computing Corp.
MBS Communications Ltd.
Micom-Interlan
Microplus
Microsoft Corp.
Morino Associates Inc.
NCR Computer Inc.
Novell Inc.
Pacer Software Inc.
Point-4 Data Corp.
The Protocol Team Inc.
Quantum Software Systems Ltd.
Rolm Mil-spec Computers
S&H Computer Systems Inc.
Santa Cruz Operation Inc.
Sun Microsystems Inc.
Syntax Systems Inc.
Sytek Inc.
Tandy Corp.
Tangent Technologies Inc.
TOPS Computer
Touch Communications Inc.
Torus Systems
Ungermann-Bass Inc.
Unisys
Univation
Venturcom Inc.
Wang Laboratories Inc.
Waterloo Microsystems Inc.
Western Digital Corp.

---

mechanisms to allow user application programs to converse, synchronize, and coordinate their activities. Features such as remote procedure calling, or message passing, or invocation mechanisms would be found here. Details of a few mechanisms will be found in Chapter 6.

Table A-11 lists vendors who indicate in their literature that they provide some form of distributed data processing software.

Table A-11. DDP Network Software

American International Communications Corp.	Netec International Inc.
BB&N	Network Solutions
CMC/Communications Solution Inc.	Network Systems Corp.
Concept Development Systems Inc.	Northern Telecom Inc.
Contel	Novell Inc.
Digital Equipment Corp.	Orion Network Systems Inc.
Datanex	The Pande Group Inc.
Datapoint Corp.	Peregrine Systems Inc.
Ericsson Information Ab.	The Protocol Team Inc.
Harris Corp.	RCA Global Communications Inc.
Hewlett-Packard Co. Business Systems Sector	Software AG of North America
Honeywell Bull Inc.	Sun Microsystems Inc.
IBM	Sydney Development Corp.
Intel Corp Systems Group	Sydney Ltd.
Levi Ray & Shoup Inc.	Syntax
Linkware Corp.	The Systems Center Inc.
NCR Comten Inc.	Tandy Corp/Radio Shack
	Texas Instruments

MODEL/SIMULATE/FORECAST/PLAN

This category of software covers a wide spectrum of networking software. The emphasis of software in this category is to provide services that allow users to model a LANs structure, to simulate its performance so as to aid in LAN selection or management, or to provide basic simulation services that allow for distributed simulations to be written and executed. This would allow modelers to write much more comprehensive and complete models and have the added resources to run such simulations in realistic time frames. The distributed system provides added CPUs that can be utilized to perform the many jobs a simulation must perform. These added CPUs provide added CPU cycles that would otherwise be unavailable to a modeler and therefore would preclude him or her from performing some tasks.

Other software lumped into this broad category includes forecasting and planning software. These include software to manage businesses, to plan schedules, to forecast earnings, or any other aspect of an enterprise's operations. As networks mature and better tools arise, this category of LAN software will expand. For now, I will mention only what I am familiar with and leave the future up to LAN software developers.

Table A-12. LAN Modeling/Simulation Tool Vendors

Acks Computer Applications, Inc.	The Info Group
Architecture Technology Corp.	Integrated Network Corp.
Aries Group Inc.	Intel Corp Systems Group
AT&T Communications	John Bridges and Assoc.
BBN Communications Corp.	LAN Services Inc.
BGS Systems	Netcomm Inc.
CACI Inc.—Federal	Network Synergies Inc.
CHLAMTAC	The Pande Group Inc.
Communications Facilities and Services Inc.	Performix Software Corp.
	Pritsker and Assoc.
Connections Telecommunications Inc.	The Protocol Team Inc.
Contel Business Networks	Quintessential Solutions
Contel ASC	Sargent Fiber Optic Services
Data Transmission Essentials Inc.	Siemens Data Switching Systems Inc.
DMW Commerical Systems	Technetronic Inc.
Economics and Technology Corp.	Telco Research
4-Degree Consulting	Telecomsyst Services Inc.
General Network Corp.	Unisys
Hughes Network Systems	Vector Software
HTL Telemanagement Ltd.	Versa-lite Systems Inc.
IBM	

Model/Simulate

Most software available today in this category falls under the simulation of a LAN's sphere. That is to say, the typical package deals with the simulation of LANs rather than providing tools to allow distributed simulations over LANs to be built and used. Simulation-language vendors queried indicate that much research is ongoing to provide such simulation languages, but they are still a little way away from fully functional products. Therefore, the list provided in Table A-12 contains vendors who indicate that they have tools to aid in LAN development or a simulation, or have basic capabilities for LAN modeling or simulation.

Forecast/Plan

Forecast and planning software come in many flavors. For example, forecasting could be projecting sales or inventory or production for a company. Software for such functions must have mechanisms to take present-day information on production, consumption, stockpiling, and use this to project the future. Likewise, planning software may be used to create schedules, or plan for raw material requirements to meet the future needs of production, for example. Typical of this type of software are presently available business-management/accounting packages. These typically provide mechanisms to collect data, correlate/process data, and construct various representations of the data for use in forecasting, planning, and managing data.

Table A-13 lists vendors who produced business/accounting software packages. Details of a few will be provided in Chapter 6.

DEVELOP/MAINTAIN/EXECUTE SOFTWARE

This categorization of LAN software addresses user needs for developing their own LAN applications programs. As such, embodied in this area are software engineering environments, as well as user-friendly environments, and code development tools such as editors, linkers, loaders, debuggers, and compilers. The list goes on and on.

Languages for LAN applications have a way to go, although there is much research ongoing. Most take a more traditional approach by using languages that already exist along with extensions to allow communications and synchronization to be accomplished. Table A-14 lists vendors who list software development tools and environments supportive of the LAN environment.

CAD/CAM SOFTWARE

Computer-aided design and computer-aided manufacturing are the buzz words of the business sector at large. These new products are aimed at increasing designers' productivity and accuracy and manufacturers' productivity and quality. The main basis of each is to extract time-consuming tasks from present processes and automate them, utilizing computer technology. In the computer-aided design area, this automation includes performing tasks such as version control, freehand drawing, measurements, auto scaling for all components, 3-D notations for design perusal, simulation of operators, or other design factors. The ability to select precanned symbols and fonts, to produce bill of materials, cost estimates, curve fitting, layering of drawings, pan and zoom, and a myriad of other features.

Computer-aided manufacturing deals with a wide range of operations from quality control inspection of parts to actual assembly. The computer has been installed throughout manufacturing installations into a variety of positions; however, in this section we will deal more with aiding setup as the manufacturing aspect, leaving actual factory control to the latter sections of this chapter.

Vendors who indicate that they provide some CAD/CAM capability and who indicate it can be used to transmit and shape designs among users are listed in Table

Table A-13. Forecasting/Planning

Access Technology	JMI Software Consultants Inc.
Adelie Corp.	Kurta Corp.
Advanced Computer Systems	LCS Telegraphics
Advanced Computer Technology	Lex Computer Systems
Advent Software Inc.	Lifeboat Associates
Ashton-Tate	Lotus Corp.
AT&T Business Markets Group	MAI Basic Four Inc.
Boeing Computer Services Co.	Martin Marietta Data Systems
Brain Power Inc.	MCS Computer Products Inc.
BTI Computer Systems	Micro Systems Software
Complete Computer Systems	Microvitec Inc.
CompuLan Corp.	Morino Associates
Compuscan Inc.	McDorf Comp RG
Computer Extension Systems Inc.	Northern Telecom Inc.
Continental Resources Inc.	Pansophic Systems Inc.
CYMA/McGraw-Hill	Peregrine Systems Inc.
Data Rentals/Sales Inc./Data General	Philips International BV
Digital Equipment Corp.	Quebec Telephone
Electronic Interface Assoc Inc.	Quinn Essentials Inc.
ERI Electro Rep Datacomm Products Inc.	The Santa Cruz Operation Inc.
Ericsson Information Systems AB/ESCA Inc.	Snow Software
Execucom Systems Corp.	Software AG of North America
JR Fall & Co.	South Central Bell Advanced Systems
GTE Data Services Inc.	Stonehouse & Co.
Hawkeye Grafix Inc.	Syntrex Inc.
Hewlett-Packard Co. Business Systems Sector	Tandy Corp./Radio Shack
Honeywell Bull Inc.	Tele-Data Advanced Information Sys Inc.
IBM	Texas Instruments
The Info Group	Trax Softworks Inc.
Information Dimensions Inc.	TSI
Innovative Software Inc.	Unisys Corp.
Interactive Information Systems Inc.	Wang Laboratories Inc.
Intertec Diversified Systems Inc.	Xiox Corp.
	Xtend Communications

A-15. Details of a few packages were seen in Chapter 6. The emphasis from a LAN environment is that users can share data from a design, operate on this data, and recombine additions, deletions, and augmentations into a global (albeit not necessarily centralized) database for the designs. True integrated CAD/CAM packages for large designs using multiple machines are here, and they are expanding rapidly.

Table A-14. Applications Development, Maintenance, and Run-Time Support Software Vendors

Advanced Business Microsystems Inc.	IBM Informix Software Inc.
Apollo Computer Inc.	Innovative Software Inc.
ArtSoft Inc.	Lotus Development Corp.
AT&T	Microsoft Inc.
Banyan Systems Inc.	Motorola Inc.
Bartel Software Inc.	Ionet Communications
Basis Inc.	Nostradamus Inc.
Blaise Computing Inc.	Novell Inc.
Blyth Software	Prime Computer
Borland International	Project X Software Development
CBIS Inc.	Quarterdeck
Central Point Software Inc.	SoftLogic Solutions Inc.
3-Com Inc.	Software Group
Clarion Software	Starlight Software
Control Data Corp.	Sun Microsystems Co.
Corvus Systems Inc.	Torvus Systems Inc.
Data General Corp.	Unisys
Digital Equipment Corp.	Univation
Futuresoft	Wang Laboratories Inc.
Hewlett-Packard Co.	Western Digital Corp.
Honeywell Bull Inc.	

COMPUTER-AIDED PUBLISHING

Another aspect of computer-aided design software is computer-aided publishing, or electronic publishing. This is a form of CAD/CAM, where the computer is used in all aspects of the development and production processes. As a matter of fact, this entire book was developed using CAD/CAM tools to write the text, produce the figures, format them into the form now seen, produce the print plates, print the books, bind them, and package them for shipment. The computer acted in all phases to speed up the process of producing these books. The CAD tools provide the book writer/editor/publisher with the ability to design multiple ways to convey information, examine its appearance quickly, and decide on the best presentation, without reverting to pen and pencil designs. The media of creation and innovation has and will continue to provide a fabulous means to produce the written word.

Publishing CAD utilizes tools such as word processors, graphics design, version control systems, text manipulators and formatters, spell checkers, style checkers, print descriptors, and numerous other tools of the trade. Table A-16 lists vendors who

Table A-15. Engineering/Design (CAD/CAM) Software Vendors

Accel Technologies	Harris Corp.
Adage Inc.	Hewlett-Packard
American Small Business Computers Inc.	Hewlett-Packard Co. Business Systems Sector
AT&T	Honeywell Bull Inc.
Autodestrine	Intergraph Corp.
CalComp	Kurta Corp.
Communications Satellite Corp.	LCS Telegraphics
Contel	Lifeboat Associates
Continental Resources Inc.	Marconi Instruments Ltd.
Creare Inc.	Micrografx
Data/Com South Inc.	ROR Holdings
Data I/O Corp.	Scan-Graphics Inc.
Data Rentals/Sales Inc.	Shirley Institute
Data Technology Inc.	Signal Technology Inc.
Euro-Bit Spa	Sky Computers Inc.
Evans & Sutherland Computer Corp.	Society of Automotive Engineers Inc.
Eyedentify Inc.	Tandy Corp/Radio Shack
Foresight Resources Corp.	Tektronix Inc.
General Digital Corp.	Unisys Corp.
General Network Corp.	Valid Logic Systems Inc.
Gould Inc. Semiconductor Div./Graftec Inc/Unisys	Venturcom Inc.
Gulf Publishing Co.	

indicate that they provide CAD/CAM systems for publishing documents. A representative system was reviewed in Chapter 6.

CONTROL SOFTWARE

This category is sort of a catch-all. Included in this category are LAN management software, security, access method software, gateway, and bridge software, as well as factory automation software. The goal is to list software that provides control over a LAN activity. For example, the management software for LANs allows users to collect statistics on use, define problem areas, allocate assets, and do numerous other LAN control functions. Security and LAN access deal with controlling who accesses a LAN and how they do it. This type of control is imperative in your competitive LAN environments. LAN software to access other LANs in the form of gateways and bridges control access between networks.

Table A-16. Publishing CAD/CAM Software Vendors

- American Research Corp.
- Atari
- Adobe Systems Inc.
- Aldos Corp.
- Bitstream Inc.
- IMSI
- Lanier
- Manhattan Graphics
- Qandu Computing Inc.
- T/Maker

The final category, factory automation, collects together LANs that provide control software for automated factory equipment control.

Table A-17 covers vendors who produce management software. Table A-18 lists vendors who produce security and LAN/WAN access software. Table A-19 covers vendors who provide bridge/gateway software, and Table A-20 lists vendors who produce factory automation/control software.

Table A-17. LAN Management Software Vendors

AT&T	
Atlantic Research	Network General
Avant-Garde Computing	Paradyne
Bridge Communications Inc.	Racal-Milgo
Case Communications Inc.	Racal-Vadic
Codex	Retix
CimCom	Symplex
3-Com	Teleprocessing products
Data Com	Timeplex
Datatec	Waterloo Microsystems Inc.
Digilog	West Electronics Ltd.
Digital Communications Assoc.	Western Digital Corp.
Dynatech	Winnertech Corp.
Excelan	Winterhalter Inc.
General Data Com	Woolf Software Systems Inc.
Infinet	Xecom Inc.
Infotron Systems Corp.	Xerox Corp.
IBM	Xicom Technologies Corp.
Microsoft	Xios Systems Corp.
Network Systems	XTEND Communications

Table A-18. Security Packages Vendors

File Security Software:

Ardy Produckter AB Ardy Electronik
Artisoft Inc.
Cennoid Technologies Inc.
Codercard Inc.
Digital Communications Assoc.
Digital Pathways Inc.
Frontier Technologies
General Digital Corp.
Goal Systems Intl.
Hawkeye Grafix Inc.
Microstar Software Ltd.
Norton-Lambert Corp.
Novell Inc.
Prime Factors Inc.
Rusco Electronics Systems
Simpact Assoc Inc.
Software AG of North America
Sytek Inc.
Talon Technology
Tandy Corp./Radio Shack
Technical Communications Corp.
TSI
Unisys
Western Digital Corp.
Winterhalter Inc.

Network Security Software:

American Intl. Communications Corp.
Ardy Produkter AB Ardy Electronik
Avant-Garde Computing Inc.
BBN Communications Corp.
Cincom Systems Inc.
Codercard Inc.
Computer Security Institute
Digital Communications Assoc.
Digital Pathways Inc.
Duquesne Systems
EDP Security Inc.
Frontier Technologies
General Digital Corp.
LanQuest Group
LeeMah DataCom Security Corp.

(Continued)

Table A-18. *Continued*

Novell Inc.
On-Line Software Intl. Inc.
Prime Factors Inc.
Racal-Milgo
Simpact Assoc. Inc.
Sytek Inc.
Tandy Corp./Radio Shack
Technological Software Concepts
Technology Concepts Inc.
Telcor Systems Corp.
Telemetrix Inc.
Telexpress Inc.
Telindus NV
Ungermann-Bass Inc.
Western Digital Corp.

Telecommunications Access Method Software:

Account-A-Call
Agile Systems Inc.
Arts Computer Products Inc.
Bell Communications Research
Bizcomp Corp.
Computas Communications
Concept Development Systems Inc.
Consumers Software Inc.
Crosstalk Communications/DCA
Datanex Inc.
Digital Management Group
Duquesne Systems
Evans Griffiths & Hart
Evergreen Consulting Inc.
Gateway Communications Inc.
Gateway Microsystems Inc.
General Digital Corp.
Harris Corp.
Hawkeye Grafix Inc.
Hughes Network Systems
Jeumont Schneider
Lex Computer Systems
Micro-Integration Corp.
Microcom Inc.
Modern Technology Inc.
NCR Comten Inc.
Network Solutions

(Continued)

Table A-18. *Continued*

Northern Telecom Inc.
Norton-Lambert Corp.
Omzig Corp.
PC Communications Ltd.
Popular Programs Inc.
The Protocol Team Inc.
RCA Global Communications Inc.
San/Bar Corp.
Satellite Technology Management Inc.
Sydney Development Corp.
Syscom Inc.
Systems Compatibility Corp.
Sytek Inc.
Talon Technology
Tandy Corp./Radio Shack
Tele-Data Advanced Information Sys Inc.
Telelogic Inc.
Telemetrix Inc.
Trans-Lux Corp.
Unisys
Vector International
West Electronics Ltd.
XTEND Communications

Table A-19. Network Interface/Gateway Software (X.25, SNA, DECNET)

Adax Inc.	Computas Communication
American Teleprocessing Corp.	Comterm Inc.
Applied Communications Inc.	Contel
Banyan Systems Inc.	Convergent Technologies
BBN Communications Corp.	CXI Inc., a Novell Co./Data Architects Inc.
Bell Communications Research	Data Interface Systems Corp.
Boeing Computer Services Co.	Data Rentals/Sales Inc.
Brown's Operating System Services Ltd.	A Datacom Solution Inc.
Cambridge Computer Associates Inc.	Datanex Inc.
Channel Systems Intl Inc.	Datapoint Corp.
Chi Corp.	Digital Communications Assoc.
Cleo Software	Digitech Industries Inc.
CMC	Dove Computer Corp.
CMQ Communications Inc.	Duquesne Systems.
Comm Pro Associates	Eastman Communications
Communications Solution Inc.	

(Continued)

Table A-19. *Continued*

Eicon Technology Corp.
ETE Group Spa
Evergreen Consulting Inc.
FlexLink Intl. Corp.
Forest Computer Inc.
Fox Research
Frontier Technologies
Gateway Communications Inc.
Gateway Microsystems Inc.
Gcom Inc.
Halley Systems Inc.
Lanier Business Products/Sector of Harris Corp.
Harris Corp.
Hewlett-Packard
Hewlett-Packard Information Networks Div.
Honeywell Bull
Hughes Network Systems
Ideassociates Inc.
Information Technologies Inc.
InnoSys Inc.
Ins Inc., an ICOT Co.
Intel Corp. Systems Group/Intelligent Technologies Intl Corp.
Interlink Computer Sciences
Jeumont Schneider
JMI Software Consultants Inc.
Jupiter Technology Inc.
Levi Ray & Shoup Inc.
MAI Basic Four Inc.
Marconi Instruments Ltd.
MBS Communications Ltd.
McDonnell Douglas Info Systems Intl.
Memotec Data Inc.
Meridian Technology Corp.
Metacomp Inc.
Microtronix Systems Ltd.
Modcomp/NCR Comten Inc.
Netserv/Network Research Corp.
Network Software Assoc.
Network Solutions
Nokia Information Systems
Northern Telecom Inc.
Norton-Lambert Corp.
Novell Inc.
Omnicom Inc.
Orion Network Systems Inc.
OST Inc.
The Pande Group Inc.
Pathway
Proteon Inc.
The Protocol Team Inc.
Rabbit Software Corp.
Retix/S-Com Computer Systems Engrs Ltd.
The Santa Cruz Operation Inc.
Satellite Technology Management Inc.
SBE Inc.
Siemens Data Switching Systems Inc.
Simpact Assoc. Inc.
Software Support Intl. Inc.
Spartacus Inc., a Fibronics Co.
Sydney Development Corp.
Sync Research
Sync Solutions Inc.
Syntax
System Strategies Inc.
The Systems Center Inc.
Sytek Inc.
Talon Technology
Tandy Corp./Radio Shack
Techland Systems, Bluelynx Div. of MicroIntegration
Technology Concepts Inc.
Telefile Inc.
Telematics International Inc.
Telemetrix Inc.
Tellabs Inc.
3Com Corp.
Til Systems Ltd.
Titn Inc.
Topcode
Travtech Inc.
Ultranet Ltd.
Ungermann-Bass Inc.
Unisys Corp.
URGEO Software Inc.
Vector International
Vitalink Communications Corp.

(Continued)

Table A-19. *Continued*

Wang Laboratories Inc.	Xicom Technologies Corp.
Waterloo Microsystems Inc.	Xiox Systems Corp.
Wellfleet Communications Inc.	Xmit AG
Western Digital Corp.	Young and Associates
Winterhalter Inc.	

SUMMARY

This appendix provided a glimpse of the volume of vendors who supply software to support networking. The reader who is interested in details is directed to Chapter 6 and to the specific vendor's product descriptions. I apologize if any vendor is left out or if any errors in listings have occurred. Please bring such discrepancies to the publisher's attention so that corrections/additions can be made.

Table A-20. Factory Automation Software

Ardy Produckter AB Ardy Electronik	
The Automation Group Inc.	Information Dimensions Inc.
Caere Corp.	Interactive Information Systems Inc.
Complete Computer Systems	Intertec Diversified Systems Inc.
Computer Extension Systems Inc.	JMI Software Consultants Inc.
Contel	MAI Basic Four Inc.
Continental Resources Inc.	Marcam Corp.
Data Translation Inc.	Marconi Instruments Ltd.
Data World Products Inc.	Martin Marietta Data Systems
DSP Technology Inc.	Modcomp/The Pande Group Inc.
Frontier Technologies	Simpact Assoc. Inc./STSC Inc.
GR Electronics Ltd.	Sydney Development Corp.
Hewlett-Packard Co. Business Systems Sector	Tandy Corp./Radio Shack
	Telettra Spa
Honeywell Bull Inc.	TSI
Industrial Computer Designs Inc.	Ungermann-Bass Inc.
Infolink Corp.	Unisys Corp.

REFERENCES

Anderson, E. and E. Douglas Jensen, "Computer Interconnection Structures: Taxonomy, Characteristics, and Examples," *Computing Surveys*, December 4, 1975.

Anderson, Lee. *Fault Tolerance Principles and Practice.* Prentice-Hall, International, 1981.

ANSI X3.66-1979. Advanced Data Communications Control Procedures.

Charette, B. and Stokenburg, Wallace. *Unified Methodology for Developing Systems.* McGraw-Hill, NY 1986.

Cheriton. The Thoth System: Multiprocess Structure and Portability. Elsevier Pub., New York, 1982.

Clark, D. "An Introduction to Local Area Networks," *Proceedings of IEEE* Vol. 66 No. 11, November 1978.

Davis, C. G. and Vick, C. "The Software Development System." In *IEEE Transaction on Software Engineering* Vol. SE-3, No. 1, Jan. 1977.

Date. *An Introduction to Database Systems.* Addison Wesley, Reading, MA, 1983.

Demarco, T. *Structures Analysis and Systems Specification,* Prentice-Hall Englewood Cliffs, NJ 1979.

DesJardins, Richard. "ISO ANSI Reference Model of Open Systems Interconnection." In *Proceedings Trends and Applications: 1980—Computer Network Protocols* (National Bureau of Standards, Gaithersburg, Maryland), May 1980, pp. 47–58.

Desrochers. *Principles of Parallel and Multiprocessing.* Intertext Publications and McGraw-Hill, New York, 1987.

Desrochers and Paul J. Fortier. *Modeling and Analysis of Local Area Networks.* Multiscience Press and CRC Press, 1990.

DIS 8973. Open Systems Interconnection—Connection-Oriented Transport Protocol Specification.

Fortier, Paul J., "Generalized Simulation Model for Evaluation of Local Computer Networks," HICSS, 1983.

Fortier, Paul J., "Survey of Local Area Networks and Related Topics," Naval Underwater Systems Center, Newport Lab, Newport, RI May 1980 (Unclassified).

Fortier, Paul J., and Leary, Richard G. "Software Simulation Study of Local Computer Networks," Naval Underwater Systems Center, Newport Lab, Newport, RI May 1980 (Unclassified).

Fortier, Paul J. "A Communications Environment for Real-time Distributed Control," ACM Northern Regional Conference, March, 1984.

Fortier, Paul J. *Design of Distributed Operating Systems Concept and Tools.* Intertext Publications and McGraw-Hill 1986.

Fortier, Paul J. *Handbook of Local Area Network Technology.* Intertext Publications and McGraw-Hill, 1986.

IEEE 802.2—1985. The Institute of Electrical and Electronics Engineers. *Logical Link Control.* American National Standard ANSI/IEEE Std 802.2, 1985.

IEEE 802.3—1985. The Institute of Electrical and Electronics Engineers. *Carrier Sense Multiple Access with Collision Detection (CSMA/CD) Access Method and Physical Layer Specifications.* American National Standard ANSI/IEEE Std 802.3, 1985.

IEEE 802.4—1985. The Institute of Electrical and Electronics Engineers. *Token-Passing Bus Access Method and Physical Layer Specifications.* American National Standard ANSI/IEEE Std 802.4, 1985.

Jensen, E.D. "The Honeywell Experimental Distributed Processor—An Overview." In *Computer*, Vol. II, January 1978, pp. 28–38.

Hoare. "Monitors: An Operating System Structuring Concept," *Comm of ACM* 17(10) 1974.

Kobayashi, A. *Modeling and Analysis: An Introduction to Systems Performance Evaluation Methodology.* Addison-Wesley, Reading, MA, 1980.

Komoda, W., et al. "An Autonomous, Decentralized Control System For Factory Automation." In *IEEE, Computer*, Vol 17, No. 12, Dec. 1984.

Metcalfe, R.M., and Boggs, D.R. "Ethernet: Distributed Packet Switching for Local Computer Networks." In *Comm. ACM*, July 1976, pp. 395–404.

Mier, E. E. "Who's on Board for Ethernet?" In *Data Communications*, May 1984, pp. 46–47, 62.

MIL-STD 1778, Transmission Control Protocol.

Peterson, W., and Weldon, E. *Error Detection Codes.* MIT Press, Cambridge, MA 1972

Robinson. *The HDM Handbook* Volume I. "The Foundations of HDM." SRI Report #4828 SRI INT 1979.

Sauer, C., and K. M. Chandy. *Computer Systems Performance Modeling.* Prentice-Hall, Englewood Cliffs, NJ, 1981.

Schwartz, M. *Computer Communications Network Design and Analysis.* Prentice-Hall, Englewood Cliffs, NJ, 1977.

Sha. "Modular Concurring Control and Failure Recoup—Consistency, Correctness, Optimization," Ph. D. thesis, ????, 1983.

Shoch, J. F. *An Annotated Bibliography of Local Computer Networks.* Xerox Corp. Rpt., 3rd Edition, April, 1978.

Shoch, J. F., et al. "Evolution of the Ethernet Local Computer Network. In *Computer*, August 1982. pp. 10–26.

Stallings, W. *Data and Computer Communications, Second Ed.* Macmillan, New York, 1988.

Tannenbaum, A. *Computer Networks.* Prentice-Hall, Englewood Cliffs, NJ 1981.

Tokuda, H. "Shoshin, a Distributed Software Testbed." Ph. D. thesis, University of Waterloo, Waterloo, Ontario, Canada, 1983.

Tsay. "Mike: A Network Operating System for the Distributed Double-loop Computer Network." Ph. D. thesis, Ohio State, 1981.

Wasserman, A. "Use and Methodology for the Design and Development of Interactive Information Systems in Formal Models and Practical Tools for Information Systems Design," ed. H. Schneider, North Holland 1979.

Wittie, Larry D. "Architectures for Large Networks of Microcomputers." In *Proceedings of the Workshop on Interconnection Networks.* Purdue University, West Lafayette, Indiana, April 1980, pp. 31–40.

Xerox Corporation, Ethernet Specification, September 1980.

Xerox Corporation, Internet Transport Protocols, December 1981.

Xerox Corporation, Ethernet Specification, Revision 2.0, December 1982.

INDEX